SOCIETY FOR THE STUDY OF HUMAN BIOLOGY

SYMPOSIUM SERIES: 28

T0292359

Human mating patterns

PUBLISHED SYMPOSIA OF THE

SOCIETY FOR THE STUDY OF HUMAN BIOLOGY

Numbers 1–9 were published by Pergamon Press, Headington Hill Hall, Headington, Oxford OX3 0BY. Numbers 10–24 were published by Taylor & Francis Ltd, 10–14 Macklin Street, London WC2B 5NF. Further details and prices of back-list numbers are available from the Secretary of the Society for the Study of Human Biology.

Human mating patterns

Edited by

C. G. N. MASCIE-TAYLOR

Department of Biological Anthropology, University of Cambridge

A. J. BOYCE

Department of Biological Anthropology, University of Oxford

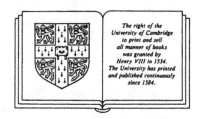

The right of the
University of Cambridge
to print and sell
all manner of books
was granted by
Henry VIII in 1534.
The University has printed
and published continuously
since 1584.

CAMBRIDGE UNIVERSITY PRESS
Cambridge
New York   New Rochelle
Melbourne   Sydney

CAMBRIDGE UNIVERSITY PRESS
Cambridge, New York, Melbourne, Madrid, Cape Town, Singapore, São Paulo, Delhi

Cambridge University Press
The Edinburgh Building, Cambridge CB2 8RU, UK

Published in the United States of America by Cambridge University Press, New York

www.cambridge.org
Information on this title: www.cambridge.org/9780521114684

First published 1988
This digitally printed version 2009

A catalogue record for this publication is available from the British Library

Library of Congress Cataloguing in Publication data

Human mating patterns
edited by C.G.N. Mascie-Taylor, A.J. Boyce
(Society for the Study of Human Biology symposium series ; 28)
Includes index.
1. Human reproduction - Congresses.
2. Sex customs - Congresses.
3. Human population genetics - Congresses.
I. Mascie-Taylor, C.G.N.   II. Boyce, A.J. (Anthony J.)   III. Series
GN253.H86 1989 304.6'3 - dc19 88 - 27534

ISBN 978-0-521-33432-7 hardback
ISBN 978-0-521-11468-4 paperback

CONTENTS

Social, religious and cultural factors

PREFACE

Mating Patterns was chosen as the theme for the two-day meeting of the Society for the Study of Human Biology held at the Pauling Centre for Human Sciences, University of Oxford, in April 1986. Mating patterns are important in human populations in many ways - influencing their demography, maintaining their genetic structure, affecting their genetic constitution and its rate of change, providing a mechanism for social and cultural cohesion, among others. Human biologists, therefore, examine the role of mating patterns from a number of different but interrelated perspectives and the programme was designed to encompass these various views.

The first session was concerned with historical and demographic aspects of mating patterns. This was followed by three papers reviewing mate choice and assortative mating. On the second day the sessions dealt with the medical and biological aspects of inbreeding and the meeting concluded with reviews of social, religious and cultural aspects of the subject.

We would like to thank the many people who helped with the Symposium, including the Chairmen, Professor Derek Roberts and Professor Geoffrey Harrison. Additional financial support for the Symposium was provided by The Royal Society, to whom we express our thanks.

C. G. N. MASCIE-TAYLOR
Department of Biological Anthropology
University of Cambridge, England

A. J. BOYCE
Department of Biological Anthropology
University of Oxford, England

PART I

HISTORICAL AND DEMOGRAPHIC STUDIES

MATING PATTERNS - AN HISTORICAL PERSPECTIVE

A. D. J. MACFARLANE

Department of Social Anthropology,
University of Cambridge, Cambridge, U.K.

There are a number of interlinked puzzles related to the pattern of mating. Currently, most modern and a growing number of developing societies exhibit a homeostatic demographic regime where the balance of population to resources is maintained through fertility control. Fecundity or natural fertility is kept well in check through relatively late age at marriage, non-marriage, contraception and abortion. These preventive checks, to use Malthus' distinction, allow rational, conscious and artificial control over population. Marriage is based on personal choice, the chief pressures being the psychological and economic state of the individual. The relatively low birth rate makes possible a relatively low death rate. The human body is controlled by individual mind and individual feelings. This accords with Levi-Strauss's (1949) 'complex' marriage system, based on psychology and economics. It has not always been so, and indeed it is arguable that this demographic and marital pattern is very unusual both in time and space.

The majority of tribal and peasant societies in the past have had an 'elementary' marriage system in Levi-Straussian terms. That is to say, marriage was not based on individual but on group choice and was determined by birth status, in other words kinship position. Marriage has characteristically occurred at a very early age for women and maximum fertility was aimed for. This very high fertility was balanced by heavy mortality, either perennially or in periodic crises, often triggered by war. Thus the checks were mainly of a positive kind, acting through the biology of disease or starvation. In this classic or crisis demographic world (Macfarlane, 1976), man was at the mercy of the environment. There were periods of disturbance of the balance with rapid population growth for short periods before the positive checks operated again. When the situation now in Europe is compared with that in the great historic civilisations of India, China, Egypt or much of Europe up to the end of the eighteenth century, it is clear that a revolution has occurred. The demographic pattern is entirely different and so is the mating pattern. How and why this transformation occurred has important implications for the origins of industrialisation and the current demographic patterns in the Third World.

The study of mating patterns in the past has been transformed over the last twenty or so years by the applications of new methods and the discovery of new materials. Historical materials concerning marriages, births and deaths are extremely difficult to use and for a long time it seemed unlikely that much could be learnt in detail concerning such intimate matters before the nineteenth century. The work of historical demographers, particularly in France and England, has changed the situation. Applying the method of 'family reconstitution', that is the linking of baptisms, marriages and burials, to the registers, and combining these with listings of inhabitants and other documents, has provided a new picture of the emergence of that 'unique west European marriage pattern' to which Hajnal (1965) drew attention some twenty years ago. This study concentrates on the English phenomenon, for it was in England that it was shown in its most extreme and most precocious form.

The particular puzzle in England was to trace the connection between population movements and the origins of the industrial revolution. It was obvious that England's unique and first industrialisation could not have occurred without a particular and unusual demographic pattern. The crucial period of wealth accumulation which formed the substructure for industrialisation during the period between about 1620 and 1720 was one in which the population grew hardly at all. The static population needed to be explained. Then, with increasing force from the 1750's, population grew rapidly just as it was needed to provide the labour for industrialisation. If England had not had almost the lowest population growth rate in Europe in the seventeenth and the fastest in the nineteenth, it is likely that industrialisation would not have occurred and our world would be a very different place.

For a long time the major explanation of these patterns was sought in the wrong direction. It was assumed that the major variable must be changes in mortality, in essence that the positive checks which had kept a 'traditional' society in balance up to the eighteenth century suddenly gave way. An example of this approach, one among many, is the well-known book by McKeown (1976). During the last five years, however, it has been convincingly shown by Wrigley and Schofield that the crucial variable is fertility, in other words it was the mating pattern that kept population in check, and then allowed it to grow. High age at marriage, little illegitimacy, and a high proportion never marrying were characteristic of the seventeenth century. Then in the middle of the eighteenth century the mean age at first marriage of women dropped from about twenty-six to twenty-three, illegitimacy rates rose dramatically, and the proportion of women never marrying dropped from about one quarter to one tenth. Thus these authors conclude that "about three-quarters of the acceleration in the growth rate which took place over the period is attributable to the increase in fertility brought about by changing marriage behaviour....." Or, to put it another way, "the changes which occurred in marriage and marriage-related behaviour in the course of the eighteenth century were sufficient to have raised the annual rate of growth of the population from zero to 1.26%, *even though there was no change in either mortality or age-specific marital fertility*" (Wrigley, 1981). This meant that "marriage now emerges holding the centre of the stage" (Wrigley, 1983).

Marriage strategies and regimes are the key to much of the social and economic history of the first industrial nation. They are also an important key to understanding what is happening in the crucial battle between population and resources in the developing world today. Put very simply, until ten years ago or even more recently, it looked as if the world was doomed to continued very rapid population growth which would lead to mass starvation and depletion of resources. That there is now some limited cause for optimism, as fertility drops in many small and large third-world countries, is due very considerably to changes in mating patterns. In particular, the central change from maximal to limited fertility caused by a rise in the age at marriage is spreading. There has been a significant rise in the age at marriage in many third world countries. Coale (1978) concludes that changes in marriage patterns, and particularly the rise in the age at marriage, "has been as effective in contributing to the recent reduction in the birth rate in the third world as the much more publicised spread of 'family planning'".

REASONS FOR THE MATING PATTERN

Having identified mating patterns, the demographers admit that they have merely set an unresolved puzzle for someone else. How did marriage act in this way, how old is this marital pattern in the west, what are the pressures that caused it? This paper draws attention to only a few of the features of marriage pattern which we now take for granted, and the matter is pursued further in an extended treatment of the same themes (Macfarlane, 1986).

The best model of the mating pattern was presented by Thomas Malthus in his "Essay on Population" at the start of the nineteenth century. He analysed the social groups in England in turn and showed how the preventive check worked in each. The wealthy were reluctant to marry because they could not afford to do so, for they would lose leisure and status if they married too soon. Farmers and tradesmen could not do so until well settled in a flourishing business. Wage labourers were faced with economic cost and social humiliation if they married too soon. Servants were comfortable if single, dismal if married. Malthus described a situation where marriage was not natural, automatic, arranged, but a choice, the conscious weighing of costs and benefits. He agreed with his chief opponent Godwin that "every one, possessed in the most ordinary degree of the gift of foresight, deliberates long before he engages in so momentous a transaction. He asks himself, again and again, how he shall be able to subsist the offspring of his union" (quoted in Place, 1967).

From Malthus' writings (Malthus, n.d.) can be abstracted the five features which he considered to be the essential pre-conditions for such a cost-benefit approach to mating and marriage. The most important was a strong acquisitive ethic: "the desire of bettering our condition, and the fear of making it worse ...... is the *vis medicatrix reipublicae* in politics ...... it operates as a preventive check to increase.....". This "spirit of capitalism" was only possible if people could individually gain from their actions, in other words if there was private property. "The operation of this natural check depends exclusively upon the existence of the laws of property and succession.....". Such private property would only have any meaning if the individual was safeguarded in its possession by a strong and just government which would allow people to hold on to their gains. Thus, for instance, it would only work if those who struggled successfully against the biological and sexual urges and put off mating were rewarded by a better life-style. Virtue was its own reward, but material comfort would be a bonus. This would inevitably lead to an unequal world, a ladder of wealth up which people would be encouraged to climb, down which they might very easily fall. The prizes must be powerful and widespread; there must be widespread affluence. "Throughout a very large class of people, a decided taste for the conveniences and comforts of life ...... are observed to prevail".

Malthus was, of course, writing a reply to the Utopian thinker William Godwin who believed that if equality could be established, private property abolished and vice eliminated, man would live happily ever after. Malthus pointed out that this was a delusion. The natural urges of man, and particularly those powerful biological and psychological urges to mate, the "passion between the sexes", would soon destroy such a proposed paradise. Here he elaborated his famous theory concerning the ability of human beings to multiply much faster than resources. Unchecked mating, leading to unchecked fertility, would bring down the catastrophes of the "positive checks" of war, famine and disease. The only rational way was to invoke and harness the lesser evils of greed and inequality inherent in society. In the battle against the urge to mate and procreate, the only force strong enough to win was the human desire for leisure and material affluence. Marriage patterns were the outcome, to paraphrase Levi-Strauss again, of the struggle between economic and psychological (or biological) factors. Mankind must learn to be responsible, to control his body by his mind, to think in the long-term, to treasure status and wealth rather than mating and children.

Malthus was aware that the battle was not merely between the economic and the biological. The latter was reinforced in many societies by cultural pressures. For instance, he discussed the ways that various religious systems either encouraged or discouraged maximal fertility. The aims of life in Hindu or Chinese ancestor religions included the desire for many children and particularly sons. It was one of the marks of Christianity, or at least the Protestant form, that it placed a higher value on celibacy than on marriage and that it placed little emphasis on the need for children. Its bachelor founder had set an example of a life where mating had no part.

The battle between the desire to mate and the desire for material affluence only occurs in certain kinds of societies. In many, it appears that there is no conflict. Wealth, however defined, is a consequence of having many children, not an alternative. This is a reminder that behind the argument that Malthus advanced were a number of ethnocentric assumptions about the natural state of human society in relation to marriage. The rules and aims of marriage implicit in the Malthusian vision are very similar to those which we considered to be natural and modern. Yet cross-cultural comparison shows that they are unusual, and it is important to analyse them before understanding the force of the Malthusian marriage pattern.

Malthus assumed monogamy, though most societies at his time practised polygamy. He assumed a fairly equal relationship between husband and wife, while most societies assumed male dominance and patriarchal power. He took for granted that marriages were for life, unbreakable, though most societies permitted easy divorce. He expected couples to re-marry, if one partner died, to a person of their choice, though the majority of societies either forbade remarriage at all, or made remarriage to a specific kinsman mandatory. He assumed that the young couple would live in their own house after marriage, though the majority of societies encouraged the young to live for some years with either the wife's or husband's family.   He expected there to be a fairly equal contribution to the conjugal fund, though the usual situation is for wealth to flow preponderantly from either bride or groom's family to the other side. These institutional rules, so self-evident yet comparatively so curious, were matched with equally strange beliefs about the nature of the marriage choice.

The Malthusian 'preventive check' was based on the assumption that it was the individual man and woman who would decide whom they would or would not marry.   By contrast, the vast majority of human societies at that time believed that marriage was too important a matter to be left to the couple themselves; it should be arranged by the parents or wider set of kin.   Malthus assumed that the individual could marry whomever he or she could attract.   The very elaborate rules which in the greater part of the world outside Europe dictated absolutely that an individual should marry within a certain group or category defined by kinship, geography, caste, class, religion or occupation, are nowhere evident in his analysis.   All this betrays an even deeper assumption, that marriage is a matter of choice.   Malthus believed that to marry or not to marry at all was a matter for decision by the individual concerned.   The almost universal assumption that marriage is part of the natural order, that to mate is, like eating, a necessary and automatic activity of human beings, an event like birth or death that happens to all, was a view not compatible with his scheme.   To marry or not to marry, to marry one person or another, to marry now or later, all these were the result of conscious deliberation, the outcome of the weighing up of costs and benefits which could even be reduced to a sheet of paper.

One of the best examples of such conscious accounting illuminatingly occurred in the life of Charles Darwin.   At the end of 1838, Darwin read Malthus, discovered the theory of natural selection, and drew up a balance sheet headed "This is the question" with "Marry" on one side and "Not Marry" on the other.   Having listed the costs and benefits, he proved that it was necessary to marry.   So on the reverse he wrote, "It being proved necessary to Marry. When? Soon or Late?"   The answer was soon (Macfarlane, 1986).   Darwin illustrates the Malthusian marriage pattern in practice in his own decision to mate; in relation to animal species he showed how the operation of the Malthusian 'positive checks', combined with maximum fertility, led to the survival of the fittest.   His life and his work therefore illustrate those two mating regimes which we are attempting to analyse.

Among the reasons Darwin gave for not marrying were "the expense and anxiety of children", with consequent "less money for books etc. - if many children, forced to gain one's bread".   Here he revealed an attitude which

Malthus again took for granted, namely that marriage, and particularly the rearing of children, would be economically and socially 'costly'. The whole Malthusian analysis was based on the weighing up of the advantages and disadvantages of marriage regarded from the individual viewpoint. It was assumed that mating, leading to children, brought real costs. The majority of human societies which existed in Malthus' day would not have seen an opposition between individual desire (biological and psychological forces) and individual wealth (economic and social pressures). Normally the two have run alongside each other, rather than in conflict. In most societies, it is precisely marriage, mating, and the children's labour and respect which are the consequence of such mating which are wealth. To talk of the cost of marriage, to see children as an expense and mating as likely to threaten individual prosperity was, until recently, an almost incomprehensible view. Wives and children are wealth and happiness.

The benefits of such a world where there is little conflict between biological urges and social or economic ends is obvious. Sexual and social satisfaction can be much more widespread. The cost, however, is the threat of a situation where the checks to rapid population growth are taken out of man's hands; absence of internal conflict is replaced by a world of periodic war, famine and widespread disease. Before modern contraception, Malthus posed a choice between the two. More recently, it has been possible, to some extent, to have both the gratification and the control.

HISTORICAL EVIDENCE

The marriage or mating pattern which Malthus examined and Darwin lived was widely established in England in the early nineteenth century and is now spreading over much of the world. It was unusual even within Europe at that time and practically unknown to all other and preceding civilisations. Trying to understand its causes leads to the question how old it was and from where it was descended. If it emerged in England in the middle of the eighteenth century as some have argued, then it could be seen as partly a by-product of what, in a circular way, it caused, namely the industrial and urban revolution. If its main features are present in the seventeenth century, then such an explanation has to be dropped, substituting perhaps some variant of the thesis that a bourgeois/puritan/protestant/capitalist revolution occurred in the sixteenth and early seventeenth centuries to cause it. If it is discovered that the main features go back to the fourteenth and fifteenth centuries, then it will be necessary to re-think many other matters as well as the mating pattern. The Reformation, the supposed political and constitutional revolutions of the seventeenth century, the growth of international trade from the start of the sixteenth century, none can be sufficient explanation. For then the mating patterns would have far deeper and more ancient roots, which have gradually become refined and now form the basis for marital patterns in much of the world.

In a survey of numerous kinds of historical records from 1300 to 1840 there was no evidence of a dramatic shift in the mating pattern at any point in the period (Macfarlane, 1986). For the period from 1600 to the present, the findings have been independently confirmed in a survey of another, though overlapping, set of sources (Gillis, 1985). Leaving on one side the detailed evidence, what are the major characteristics of this mating pattern and how far back can they be traced?

One factor is the fluctuating age at marriage, somehow linked to the market for labour, often rising very high and hence allowing a drop as in the eighteenth century when it caused the population to spurt. There is little evidence that this central feature of Hajnal's European marriage pattern was absent in the fourteenth and fifteenth centuries, and some evidence that it was present. It is certainly the case that women did not marry in their early or mid-teens as in many tribal and peasant societies. Likewise, it is clear that from at least the fourteenth century there was a selective marriage pattern, with large numbers of women, particularly servants, never marrying. Nor is there any evidence of a dramatic change in the rules, positive or negative, about whom one should or should not marry. No substantial evidence has yet been produced to show that there was ever a set of strong positive rules, based on kinship, as to whom one must or should marry. The negative rules were reduced at the Reformation, and have stayed unaltered since then except for the late nineteenth century allowing of marriage to deceased wife's sister. The only strong rule throughout the period was that the young couple should be independent from both sets of parents after marriage, setting up a separate, neolocal, residence. This led to those simple, nuclear, households which have been a feature of north-western Europe and particularly England from at least the fifteenth century. This independence was based on a particular form of funding. The customs of jointure and dowry, the balanced and important contribution from the individuals, nuclear families and friends, seem to have remained in essence unchanged from the fourteenth to nineteenth centuries. They ensured that people had to consider very carefully before marrying as to whether they would risk losing parental and other support if they married too soon.

The major aim of marriage, as shown in letters, diaries, advice books, poetry and many other sources, was primarily to satisfy the psychological, sexual and social needs of the individuals concerned. In the majority of societies, the prime aim is the desire to have children; marriage and mating are the means to that end. In England, it was the marriage and the mating which were the ends, children were the consequence, a by-product of the sexual union. The central importance of the actual mating was shown in the view that a marriage was not valid or binding without sexual consummation. Whereas in many societies a marriage ceases to exist if there are no children, in England sterility was not a ground for divorce. But proven sexual incapacity from the start was such a valid ground for declaring that there had never been a marriage. Throughout the period, for the vast majority of the population (the top few hundred families are often an exception) marriage was ultimately a private contract between individuals. The parents had some say, but ultimately a marriage could occur without their consent or even knowledge. On the other hand, marriage could not occur without the consent of the partners. These were very old rules, from before 1300, and lasting through to the present. They emphasised that the central feature of marriage was the conjugal relationship, the depth of feeling and shared interests of the couple. Marriage was not a bridge artificially constructed as a form of alliance with another group, in which the partners and children became the planks upon which political relations were built. It was a partnership between two independent adults who formed a new and separate unit, cemented by friendship, sex and a carefully defined sharing of resources.

This ancient system, balancing the contradictory pressures of desire for companionship and sex against the desire for wealth and social status, led to many compromises over time. These compromises, reflected in the greatest tradition of poems, plays and novels about love, marriage and mating produced in any society, are evident from Chaucer to Tennyson. The heart of the system was the deep attachment of one man to one woman, the feeling that each was incomplete without the other, most nobly expressed in the words of Shakespeare, Milton, Donne and others. Since the marriage was not bounded by formal rules which dictated whom one should marry, nor arranged by kin, it is not surprising that there existed a large and complex tradition concerning courtship. Courtships were characteristically lengthy, lasting for months or years, conducted by the couples themselves, and often fruitless or disastrous. The courtship was based on the widespread belief that marriages should, ultimately, be based on romantic love, a deep and passionate longing. This external force would grip an individual and resolve all the conflicts and indecisions, settling the equations and making it possible to come to a decision as momentous as this. The "instituted irrationality of romantic love" was clearly a central part of the mating pattern of England from at least the fourteenth century if not much earlier.

ORIGINS OF THE PATTERN

If it is correct that there was a free-floating, individualistic marriage
choice system as being characteristic of England from at least the fourteenth to
nineteenth centuries, the question arises as to what made such an unusual
pattern possible. Normally mating, for women, occurs at or soon after puberty
and continues steadily. Everyone capable of doing so mates. Fertility is at a
premium. Here, instead, there was a large-scale civilisation, still basically
agricultural and "pre-industrial", whose marital pattern flatly contradict the
norms of most other peasant societies. The pattern appears to be old, so it
cannot be the result of the urban and industrial revolutions of the eighteenth and
nineteenth centuries, or even of the supposed religious and political transforma-
tions of the sixteenth and seventeenth centuries. What could have caused or
allowed such a mating pattern to emerge?

A hint of an answer (Macfarlane, 1986) comes from Malthus himself. He
believed that the pattern he advocated could only exist in what would now be
termed a competitive, capitalist society. He argued in a way that reflected
earlier thinkers like Adam Smith and Mandeville, that the private vices of a
competitive market economy would add up to the general good of the society by
providing a force powerful enough to strain and channel unrestrained mating.
The private passions of accumulation, the instituted inequalities of an hierarchical society, the desire for material comfort, for leisure, for the snobbish
superiorities and esteem of friends, would be strong enough to save mankind
from the positive checks of war, famine and disease which would occur if the
"natural passion between the sexes" was unchecked. Put in other terms, he
argued that the mating pattern, the weighing of costs and benefits, the battle
between biology and economics, the constant striving and manoeuvring which
had, for the first time in human history, brought mankind out of a world of
periodic and dreadful crises up through the winding spiral of wealth, was the
familistic dimension of a particular economic and political system, or mode of
production, which today would be called capitalism.

Malthus was not the only one to notice the connection, obvious once
stated, between a pattern of mating and a socioeconomic formation. Any
anthropologist would expect that the mating pattern of a society would fit with
other features, the religion, the economy, the society. It is hardly surprising
that this was the case over the long "Bourgeois arch, which stretches from the
twelfth century to our own time" (Thompson, 1965). Others, too, saw the
connection. Engels in his work on the family and private property (1902) noticed
the ways in which the mating and marital system, with its obsession with choice,

free contract, desire to possess and own, fitted so well with the emergence of capitalism. "Capitalism" created a new world in which "the love match was proclaimed as a human right." The middle classes, according to Engels, grew and the new pattern became established as the emotional and family concomitant of capitalism. One of the most brilliant characterisations of the connection was made by Max Weber. He noted the central paradox whereby the most disruptive of "irrational" drives, the biological or sexual urges, were transformed, domesticated and mobilised at the heart of capitalism. As societies became more bureaucratic, more "rational", practising an ever greater division of labour, so there increased at their heart an impulsive, apparently irrational and non-capitalistic emotion at the individual level: "this boundless giving of oneself is as radical as possible in its opposition to all functionality, rationality and generality ..... the lover ..... knows himself to be freed from the cold skeleton hands of rational orders......" (Gerth & Mills, 1967). Thus rational capitalism and irrational love complement each other; the body, and particularly its reproductive mechanisms, have been brought under control, children are produced or not produced in accordance with the needs of the economy.

The irony is that, as Malthus showed, each individual is under the illusion that he acts freely, making the decisions. In fact, an invisible hand constrains him or her in such a way that the sum of the irrational, free, independent decisions seems to lead to rational acts as far as the general good is concerned. Another irony is that the nature of his or her impulsive emotions when "giving in" to love, taking off the controls to allow mating, is not in opposition, but merely another aspect of capitalist ethics. It helps to provide, through the harnessing of the biological urges, much of the excitement and activity within a capitalist society. It is not difficult to see that the "irrational passion" of love has many similarities to the "spirit of capitalism" itself, namely that desire to accumulate, to possess, to own, to entirely hold to oneself. There are many parallels between the market metaphors of purchase, contract, possession, and the powerful emotion that seizes the lover so that, as Dr. Johnson put it, finding that he is unhappy when not with the object of his desire, the individual rather rashly concludes that having such an object permanently with him will make for eternal happiness (Johnson 1810).

CONCLUSION

In the progress from the initial puzzles, as is characteristic of such research, each solution leads to further conundrums. It appears that the unique mating pattern which Malthus dissected was a powerful contributory factor in explaining the development of the first 'modern' and industrial society. Marriage and associated mating is the key to many things, lying at the intersection between the individual and society, economics and biology.   This marriage system seems to be centuries old, at least in England.   It varied, of course, by region, class, time, and these variations have necessarily had to be ignored.   But there is a discernible pattern lying behind the confusion of single decisions and the massive quantitative and qualitative materials available to the historian. For the anthropologist used to Africa, Asia or South America, the whole thing seems very strange and he is led to assume that such an unusual set of rules and aims must be both very recent and very transitory.   Instead, it is very old. Rather than being transitory, an approximation of the system, with some modifications such as easier divorce and contraception, is sweeping across the globe.   It also appears that part of the solution of the pattern lies in all the other features of the society which nursed it, in the political, economic, religious and social institutions labelled capitalist. If this is correct, one implication of this argument, not explored here, is that the capitalist system started at least a couple of centuries earlier than the usual orthodoxy allows.

Two facts may be stressed.   First, mating patterns do, indeed, hold the centre of the stage in explaining recent and current developments.   Secondly, there has been a revolution, from the pre-Malthusian, to the Malthusian pattern. When this occurred in England is still not clear, but the consequences are obvious.   The world of unrestrained fertility with high death rates as the main control can be contrasted to that where both fertility and mortality are controlled.   Counting the costs and benefits of these two approaches summarizes the choice that many societies are now experiencing.   The benefits of the Malthusian pattern are material and economic and the avoidance of the periodic horrific crises of the *ancien regime*.   The lightening of the load of women's bodies, the long period of relative freedom before marriage, the liberty to marry or not marry, the ecstasies and pleasures of romantic love and a marriage based on choice and companionship are other advantages.   The costs include the sexual and psychological frustration, at least in pre-contraceptive societies, before sexual consummation;   the anxiety of wondering whether or whom one will marry;   the loneliness of many who have never married or who have lost their companion, through death or divorce;   the increased strain on marriages which become the pivot of the whole emotional system rather than a minor part of a wider family system;   the inducement to a constantly calculative approach to human and other relationships.   Whether or not the cost outweighs the benefits, the mating pattern is changing fast but, in its deep generative rules, may have many similarities to that which was practised by our ancestors many hundreds of years ago.

REFERENCES

Coale, A.J. (1978). T. R. Malthus and the Population Trend of His Day and Ours. Encyclopaedia Britannica Lecture 1978, University of Edinburgh.

Engels, F. (1902). The Origin of the Family, Private Property and the State. Chicago.

Gerth, H.H. & Mills, C. Wright (eds.) (1967). From Max Weber: Essays in Sociology. London.

Gillis, J. (1985). For Better, for Worse: British Marriages, 1600 to the Present. New York.

Hajnal, J. (1965). The European Marriage Pattern in Perspective, In: D. V. Glass & D.E.C. Eversley (eds.), Population in History. London.

Johnson, S. (1810). Works. 12 vols. London.

Levi-Strauss, C. (1949). The Elementary Structures of Kinship, 2nd edn. London.

Macfarlane, A. (1976). Resources and population, A Study of the Gurungs of Nepal. Cambridge.

Macfarlane, A. (1986). Marriage and Love in England, Modes of Reproduction 1300-1840. Oxford.

McKeown, T. (1976). The Modern Rise of Population. London.

Malthus, T.R. (no date). An Essay on Population, 2nd edn. (2 vols.). Everyman Library, London.

Place, F. (1967). Principles of Population, New edn. London.

Thompson, E.P. (1965). The Peculiarities of the English. In: R. Milliband & J. Saville (eds.), Socialist Register. London.

Wrigley, E.A. (1981). Population History in the 1980s. Journal of Interdisciplinary History, 12, 2.

Wrigley, E.A. (1983). Growth of population in eighteenth-century England: A conundrum resolved. Past and Present, 98, Feb.1983.

MATING DISTANCE
AND HISTORICAL POPULATION STRUCTURE:
A REVIEW

A. C. SWEDLUND

University of Massachusetts, Amherst.

INTRODUCTION

In biological research those of us who choose to study humans are sometimes regarded as the 'poorer cousins' of our colleagues studying other organisms. This is, of course, because of our inability to initiate controlled experiments and to delve deeply into the structure and function of our subjects *ante mortem*. However, one very noteworthy exception to this general point is clearly the case in the study of mating patterns. While it might be considered rude, or even unscientific, for an anthropologist/geneticist to preoccupy his or herself with observation of the *act* of mating in humans, the observation of marriage formation has been elevated to a highly respectable scientific endeavour with complex methodologies; and among genealogists and ecclesiastical scholars it is regarded not only as a refined and honorable subject but in some cases even a means for possible redemption.

From the classic earlier studies on the nature of migration by such notables as Zipf (1946) and Wynne-Edwards (1962) to the extensive and comprehensive literature that exists today (e.g. Boyce, 1984; Crawford & Mielke, 1982; Swingland & Greenwood, 1983) there is probably no population of organisms better understood than humans when it comes to mating systems. This is not only the case because with humans we can observe the formation of couples and families, but also because humans have the distinct advantage of writing it down or telling us about it. These facts have prompted some investigators (e.g. Harpending & Ward, 1982) to turn the tables, so to speak, and to advise population biologists on the lessons learned from the study of human mating systems. Phenomena that are seldom, if ever, measured, and usually only contemplated for other organisms, become endless fodder for the computer-literate human biologist.

This embarrassment of riches has spawned one of the most extensively studied and reviewed subjects in human population biology (e.g. Swedlund, 1980,

1984). The present review outlines some general problems of research, discusses recent methodological and substantive contributions focussing on work that incorporates historical data, and finally summarises some recent findings from the Connecticut River Valley of Massachusetts that bear on the study of human population structure.

For human biologists the overriding issue in the study of mating distances has been, of course, their effect on the genetic structure of populations. Approximately twenty years ago there was a noticeable lack of good empirical research on the mating systems of different cultures and their resulting genetic structure. There followed a kind of 'golden age' of research on this topic, during which literally hundreds of contemporary non-western and historical European, Oriental and North American populations were studied.

Major methodological inroads were made in the study of population structure that have directed most of the subsequent research (e.g. Schull & MacCluer, 1968; Morton, 1969). Technological breakthroughs came in the form of accessible, high-speed computers that were capable of processing complex simulations, gene markers, and family surnames with considerable facility. Indeed, the bulk of those questions which were being asked twenty years ago regarding the role of gene flow, genetic drift and other demographic factors, have been answered. What remains is really fine tuning. This has prompted many genetically oriented biologists and anthropologists to move on to questions of genetic epidemiology or away from population genetics altogether and into the biochemical structure of the human genome. For those still fascinated with the complexities of human demographic behaviour and their role in both genetic and non-genetic variation, there is a plethora of work to be done, whilst acknowledging past progress and without becoming mired in replication of the accepted and the verified.

Recent volumes (Boyce, 1984; Mielke & Crawford, 1980; Crawford & Mielke, 1982) and a series of other review articles have added significantly to our understanding of some of the remaining issues. Perhaps most noteworthy of late is an increasing resolution of the effects of spatial/environmental variability and variability of migrants on genetic structure. In a previous discussion (Swedlund, 1984) attention was focussed on three dimensions of mating distance: (a) the topography, (b) the patterns of movement, and (c) the movers (migrants). The research outlined below has addressed each of these topics and answered some of the questions raised.

In the past there have been several comments to the effect that population structure analysis and mating distances should be useful for the analysis of other biosocial phenomena (e.g. Harrison & Boyce, 1972; Harpending, 1974; Swedlund, 1980, 1984; Workman et al, 1976). We are beginning to see the results of modest experimentation in this area, particularly with the relationship between population structure and disease. It appears that, at least for some subdivided populations, the mating structure and migration data are effective predictors of communicable disease experience and mortality. If the next phase in research is characterised by greater efforts at associating mating distance with other factors, the field should flourish.

DIMENSIONS OF RESEARCH

Mating distance and what?

Whether the topic of the research is genetic or non-genetic, there is a mass of data which can be utilised to correlate with mating. Most common in historical studies is probably the relatively straightforward question about the relationship between mating distances and actual geographic distances between the subdivisions of a population. Traditional procedures usually entail comparisons between matrices of geographic and mating distances by some measure of goodness of fit (see Jorde, 1980; Swedlund, 1980). Principal components or multi-dimensional scaling procedures can be used to compare the relationships between the two sets of data and, when the physical environment and the size/scale of communities are not too heterogeneous, most studies reveal strikingly good correspondence between the two. A variety of distance-decay functions have been fitted to these distributions in a relatively satisfactory way.

The more interesting research usually involves consideration of mating distance with some other set of biological or social data. Most prevalent have been studies leading to genetic inference and, traditionally, the mating distributions are compared against gene markers or other variables, commonly surname variants and quantitative anthropometric characters. Jorde (1985) has reviewed recently the study of genetic distance using gene markers; Lasker (1985) has reviewed research on surnames; and Lees and Relethford (1982) and Susanne (1984) reviewed the work on quantitative characters, and readers are referred to these studies for more comprehensive treatment. Historical studies must either have implications for the contemporary representatives of the population in question, or be satisfied with the use of surnames if estimations of genetic structure are to be made.

What is perhaps surprising in a majority of the empirical cases, even to those who stay abreast of the many regional research projects involving data on mating patterns, is how well data on mating distance and interaction tend to

predict local genetic structure. Despite all the social and environmental factors
which can militate against random mating proportional to distance, there is a
remarkable consistency between gene marker variation and levels of
heterozygosity in the subdivisions and the rates of mate exchange. Where
topography tends to be relatively homogeneous and there is not too much
hierarchy in the size or structure of communities, a simple distance/decay
function will model both the genetic variability and the mating structure more
often than not. There are several excellent reviews regarding the assumptions
necessary to infer genetic structure from migration matrices or distance-decay
models (see Jorde, 1980, 1984; Rogers & Harpending (n.d.)). Where closer
inspection of migrants suggests strongly that they are not random with respect
to a host of social, demographic and biological characteristics (e.g. Coleman,
1984; Smith, 1984; Fix, 1978, 1982; Leslie, 1985; Susanne, 1984) these
observations seldom have precise implications for the genetic structure and do
not tend to cause the gene frequency values to diverge significantly from
patterns predicted by the distance models. Cultural variability between regions
would seem to be very relevant here. Whereas, in the past, genetical
anthropologists have felt most comfortable with using the genetic data as the
dependent variable and mating or geographic distance as the independent, some
studies (e.g. Harpending & Ward, 1982) find the strength of association sufficient
and straightforward enough so as to recommend the opposite. Moreover, some
studies (e.g. Harpending & Ward, 1982; Swedlund, Anderson & Boyce, 1985) have
experimented with using the marital distance data or genetic data to recover the
real geographic distances because of their high correlation.

Scale in genetic implications

Under the broader heading of 'population structure' the scale of analyses
have ranged from subcontinental regions (e.g. Ammerman and Cavalli-Sforza,
1985; Menozzi, Piazza & Cavalli-Sforza, 1978) to national populations (e.g.
Workman et al, 1976) to the more common regionally subdivided populations and
'quasi-isolated' communities. All of these have been undertaken with varying
success, but it is generally agreed that the methodology and analysis of mating
distances, group fission and fusion, and resulting structure are most
appropriately applied at the level of regionally subdivided populations. There is
a very practical reason for this, which is that the processes best estimated by
mating distance (i.e. gene flow and gene drift), are those associated with short-
term, localised, microevolutionary processes; and it has been repeated often
(e.g. Harpending, 1974; Hiorns et al, 1977; Jorde, 1980) that these processes can
be readily swamped by sufficient levels of systematic pressure from 'outside'

migration and selection.   But, in spite of these larger processes and the difficulty in inferring the broader evolution context, the evidence from micro-evolutionary studies at the regional scale now permits reasonable confidence that mating structure and genetic structure are associated very closely in the majority of cases.

Another reassuring feature of these studies has been the fact that repeated cases have shown that marital distance tends to reflect well the results generated by the more preferred parent-offspring distance.   This can be so, even when there is relatively high variance in fertility (Jorde, 1984).

Inclusion of other variables

Recent developments in the study of mating distance and population structure have, as noted, involved the inclusion of variables other than distance/decay and in identifying a broader range of characteristics of migrants themselves. Thirdly, the shift has been towards using population structure models to address other than genetic questions. Taking these topics in order, a few observations on current research and results in Connecticut Valley are relevant.

The analyses of movement and topographic variables other than distance have tended to focus on size/density issues.   Since many genetic models assume or control size variability within subdivided populations, it is interesting to see what variability does exist and how it affects migration.   The genetic implications of size variation are discussed in greater detail in other papers (see Jakobi & Darlu, this volume).

Several studies have used marital migration data in order to analyse density dependent migration (e.g. Relethford, 1986), although Coleman (1984) warned that migration occurring at marriage may not adequately reflect lifetime mobility. It might be expected, however, that in agrarian populations which are relatively stable and sedentary, marital migration is the major form and time of movement.

In the Connecticut Valley, factors other than distance, which may have some effect were investigated, using a regression model and including as independent variables size of community, time of establishment of community, and the degree to which the communities were dependent upon agriculture or manufacturing.   The period 1830-1849 was chosen since data were available relating to twelve contiguous communities, in order to develop a multivariate model.

The intercommunity exchanges in a migration matrix were selected as the dependent variable.   The exchanges were treated as the number of individuals moving from Community i to Community j, and the number of individuals moving from j to i. In a 12 x 12 matrix this yields 132 cases (all off-diagonal cells). The independent variables included distance (the geographic distance between i and j); historical factors (the number of years since the founding date for towns i and j); environmental factors (population sizes of towns i and j, population density of towns i and j, elevation of towns i and j); economic factors (value of manufactured goods in i and j based on the State Census of 1845, value of agricultural goods in i and j based on the same census, wealth of each town adjusted for population size). Both linear and non-linear models were explored, but the non-linear model did not improve the fit significantly over a linear one. Those variables (net of distance) which achieved statistical significance, included time since founding date, value of manufactured goods and, of course, population size. Such results do pose some challenges to fundamental assumptions about migration and its presumed genetic effects.

Wood, Smouse and Long (1985) used a somewhat different model to investigate a similar problem among the Gainj and Kalam in New Guinea. Their model used a maximum likelihood procedure to measure the effect of parish size ('density'), linguistic distance, and geographic distance among 18 communities. Of particular interest to the investigators was differential migration by sex. They corroborated the Connecticut Valley findings, and those of many geographers, that size/density effects can indeed be important correlates of marriage distance. The groups studied also had high rates of endogamy and patrilocality, resulting in higher mobility for females, as is found in many agrarian populations. Swedlund (1980) states that it can probably be generalised that in sedentary agricultural populations the tendency for males to inherit land will precipitate a somewhat higher mobility rate for females.

Another exercise recently undertaken for a more general ecological survey examines the development of roads and turnpikes in the Connecticut Valley study area. This involves all 26 towns in Franklin County, whereas the marital exchange data is based only on the twelve contiguous towns for which data were available. However, the twelve towns are the northern tier of the 26 and the contrasts should be valid across both samples.

Fortunately, there were three map series available for the county, which correspond roughly to the time periods originally used for the population structure analysis. There were the 1794 (1790-1809) and 1830 (1810-1829) series maps requisitioned by the Commonwealth of Massachusetts, and the 1871 (1830-1849) Beers' Atlas maps that were drawn up privately and incorporate all counties and towns in the State. Between 1830 and 1873 there were very few roads constructed. From 1830 onwards the major features were improvements in the quality and all-season traversability of the existing routes. From these maps it is possible to track the development of the road networks over time.

The object here was to see if any estimates of genetic variables for the three time periods would conform either to the density or connectivity of road networks between the communities. Density is here defined as the number of miles of roads per square mile, and connectivity as the mean number of roads connecting any two towns in the study area.

The estimated Malecot's kinship (Figure 1) shows a noticeable change in the rate of decline with distance between periods two and three. That this corresponds to changes in the road networks is suggested by a visual inspection of the roads over the three time periods. But it is not a simple inverse relationship whereby kinship declines as road density and connectivity increase. The major transition in road development is between periods one and two, whereas the major change in kinship estimates is between two and three (Table 1). Clearly there must be some lag effect.

MATING DISTANCE AND SOCIO-ECONOMIC STRUCTURE

The analysis of mobility along lines of kinship, class and socioeconomic status has received considerable attention in the past and much of this work has been done in Britain. Non-random characteristics of migrants most often receive attention because of the potential effects of kin-structured migration (Fix, 1978, 1984).

An earlier study of Deerfield families using surnames from the marriage registrations and linking them to the tax valuation lists has suggested strongly that maintenance of an economic elite in Deerfield was based, almost from the outset, along strong kin lines (Swedlund et al, 1983). For example, surnames were isolated that appear in the marriage records before 1690 (Group A), between 1690-1723 (Group B) and after 1723 (Group C). On the basis of the tax valuation list of 1771 the population at that time was found to comprise 27% Group A surnames who controlled 57% of the wealth, 15% Group B surnames with 33% of the wealth, and 58% Group C surnames who controlled only 10% of the wealth. These relative advantages among the groups persist throughout the study period. Based on subsequent tax valuation lists and manuscript census data, Group A households controlled almost seven times as much assessed wealth as did Group C households in 1840, despite their being less well represented numerically.

Even more revealing was an analysis of 14 highly common surnames that were selected on the basis of being present after 1704 and persisting in the Town

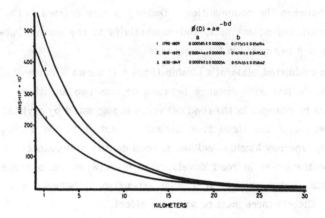

Figure 1:   The relationship between kinship (∅) and distance (in kilometers)
                for three time periods. (From Swedlund et al, 1984, by courtesy of
                Alan R. Liss, Inc.)

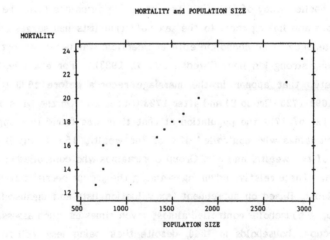

Figure 2:   The relationship between mortality rate per thousand and
                population size, 12 communities in Franklin County Massachusetts,
                1855.

Table 1:   Road density, road connectivity and genetic estimates in the Connecticut Valley, Massachusetts

Time	Road Density	Road Connectivity	Random Isonymy	F_{st}
1790–1809	0.699	0.1534	0.00325	0.00041
1810–1829	1.813	0.3700	0.00324	0.00031
1830–1849	1.841	0.4068	0.00333	0.00018

(Mean endogamy for the twelve towns is 0.7017, 0.6928, and 0.6255, respectively.)

until 1850. These names comprised mostly the Group B list noted above. By 1820 these names accounted for only 7% of the population of Deerfield, but they accounted for 54% of the assessed wealth. It is clear from these figures that time of arrival in a community is a good predictor of success.

Now, turning to marriage among these families, it is clear from reviewing the records that a large proportion of marriages occurred within and among these elite families. It was felt that there were certainly more alliances formed within the prominent families than would have been expected by chance, notwithstanding the possible scarcity of suitable mates in the early years. Using the same 14 surnames as the basis of an 'elite' group, the rate of intermarrige was measured between these families in comparison with marriages with the population at large. The results indicated that marriage within the elite families was much more common than would have been expected by chance (χ^2 = 5.6, df = 1, p <.025).

A further step was to look at isonymous marriages. The frequency of these marriages, assuming that the individuals are related affinally or consanguineally, should be a sensitive measure of wealth maintenance and consolidation. For genetic inference, it is well known that a problem arises when the assumption of monophyletism is violated and the names are the same, but of independent origin. This is also true of wealth and social-status heredity. However, a check of the Deerfield isonymous marriages using Sheldon's Genealogies (1895) revealed very little evidence of this problem. Of a total of 23 isonymous marriages, 11 were found to be between cousins of varying degrees, seven were probably related but documentation is not complete, and five showed no record beyond the marriage entry. In the case of one surname, the Jones's where one might assume no relationship because of its polyphyletic origins, four marriages occurring between 1829 and 1849 were of the same lineage and between first cousins. Thus, there seems ample evidence for the maintenance of elites through kin lines. The observed isonymy was almost 19 times the expected among elite families (Swedlund & Boyce, 1983).

It is also worth noting here that when the isonymous marriages for all time periods and all 12 communities are taken into account, slightly over 25% are exogamous marriages among the communities (Swedlund & Boyce, 1983). Virtually all of these consist of surnames/families which would not seem to violate the monophyletism assumption.

MATING PATTERNS, POPULATION STRUCTURE AND DISEASE

In recent years there has been increasing interest in applying the mating distance and population structure models to new questions in anthropology and human biology. Among these questions a fair amount of lip service has been paid to their relevance to issues in epidemiology and I will turn here to some brief observations about the study of mortality.

Several researchers are now developing a theoretical base for the analysis of population structure and disease (e.g. Fix, 1984; Harpending & Wansnider, 1982; Sattenspiel, n.d.). Harpending and Wansnider (1982), in their study of the Bushmen, suggested that the more mobile groups experience higher mortality than their sedentary neighbours. It is suggested that this results from the fact that the mobile, gathering/hunting bands are exposed to more novel pathogens as a result of moving around between new encampments and watering holes. The risk of disease and death from these pathogens is particularly high for infants and young children.

In a similar vein, Wood and Smouse (1982) analysed the density-dependent response of mortality in a group of Gainj communities in New Guinea. They found that there was, as epidemiologists have long argued, an association with density that also was most apparent in the mortality of the very young and very old. Mielke et al (1984), in their study of historical smallpox epidemics in the Åland Islands, Finland, used previously collected data on population structure to see if there were any associations between the frequency of the epidemics and the size and degree of isolation in the Åland parishes. The measure they used for size and isolation was essentially the product of population size and exogamy rate. Their results showed a strong rank correlation (r = -.918) between isolation of communities and frequency of epidemics. In subsequent analyses they have further developed the model and looked at the coincidence of epidemics in paired subdivisions as a function of their marital migration rates. In this test the inference of causality is much improved and the relationship appears to remain strong (Jorde et al, 1986).

Although several epidemics are known to have occurred among the communities analysed in the Connecticut River Valley, there is no case analogous to Åland. For one thing, by the late 18th century smallpox epidemics were rare in New England as a result of low population densities and immunisation. The epidemics that did occur tend to be of low periodicity and were more often associated with typhoid or a non-specific pneumonia/diarrhoea complex that affected mostly children. Still, it was possible to take annual rates of overall mortality and infant mortality and compare these to the population structure results. As a preliminary test, the average rate of mortality for the years 1853-1856, and the infant mortality rate for 1855 were selected as dependent variables. These were chosen not because there was any evidence of epidemics or unusual rates of mortality, but because they appeared typical and were readily available from the State Census and Annual Vital Reports. Thus, the question of interest here is whether normal mortality experience in a subdivided population can be accounted for, at least in part, by density and isolation.

The independent variables in this case are the population sizes of the twelve study communities from the 1855 State Census and the endogamy rates of the same twelve communities for the period 1830-1849. These dates are not ideal but were available from a previous study (Swedlund, Jorde & Mielke, 1984), and are thought to be representative for 1855. Regressions were carried out using size and endogamy, separately and together. The association between mortality and population size is striking. Both mortality and the infant mortality rates are highly correlated with population size ($r = .946$, and $r = .936$, respectively) and are highly significant. It is not surprising that they should both be high because infant mortality accounted for the greatest proportion of all mortality in these communities during the 19th century. The association between size and overall mortality is illustrated in Figure 2.

A multiple regression including size and endogamy was somewhat less promising. The correlation rises to $r = .952$ and the r^2 correspondingly rises a small amount, but there is no significant effect with the inclusion of endogamy and, indeed, the plot of mortality and endogamy is plagued by outliers and only a limited trend is apparent. This can be attributed easily to the small number of data points ($n = 12$) and a larger sample might establish the role of mating patterns more clearly. Marital migration data may prove to have explanatory power in this and other geographic areas and should be on the agenda for future research.

CONCLUSION

The dimensions of research outlined above and the many recent studies taking mating patterns into new directions suggest that the study of population structure is alive and well. A new body of theory is emerging (e.g. Rogers & Harpending, n.d.; Dow & Cheverud, 1985; Smouse, Long & Sokal, 1986) after several years of somewhat repetitive empirical work. While one may not be as sanguine about the future of mating distance studies as Jorde (1985) about genetic distance research, there do seem to be several interesting avenues to pursue (and/or get lost on) in the years ahead. Historical data will continue to provide some of the more interesting empirical applications. The development of research in historical epidemiology and its relationship to population structure appears to be one of the more promising applications of data on historical mating patterns.

ACKNOWLEDGMENTS

The author wishes to thank the Society for the Study of Human Biology, the Graduate Research Council (Massachusetts), and the School of Social and Behavioral Sciences (Massachusetts) for travel support. Karl Finison prepared the material on road distance and density. L. B. Jorde, R. H. Ward, A. J. Boyce and C.G.N. Mascie-Taylor provided useful suggestions on the manuscript. His thanks are also expressed to L. Sattenspiel, L. B. Jorde, A. Rogers and P. E. Smouse for making available their unpublished manuscripts.

REFERENCES

Ammerman, A. J. and Cavalli-Sforza, L. L. (1984). The Neolithic Transition and the Genetics of Populations in Europe. Princeton: Princeton University Press.

Boyce, A. J. (ed.) (1984). Migration and Mobility: Biosocial Aspects of Human Movement. London: Taylor & Francis.

Carmelli, D. and Jorde, L. B. (1982). A nonparametric distance analysis of biochemical genetic data from the Åland Islands, Finland. American Journal of Physical Anthropology, **57**, 312-340.

Coleman, D. A. (1984). Marital choice and geographical mobility, In: A. J. Boyce (ed.), Biosocial Aspects of Human Movement, pp. 19-55. London: Taylor & Francis.

Crawford, M. and Mielke, J. (eds.) (1982). Current Developments in Anthropological Genetics, Volume 2: Ecology and Population Structure. New York: Plenum.

Dow, M. M. and Cheverud, J. M. (1985). Comparison of distance matrices in studies of population structure and genetic microdifferentiation: quadratic assignment. American Journal of Physical Anthropology, **68**, 367-373.

Fix, A. G. (1978). The role of kin structured migration in genetic microdifferentiation. Annals of Human Genetics, **41**, 329-339.

Fix, A. G. (1982). Genetic structure of the Semai, In: M. Crawford & J. Mielke (eds.), Current Developments in Anthropological Genetics, Vol. 2, pp. 179-204. New York: Plenum.

Fix, A. G. (1984). Kin groups and trait groups: population structure and epidemic disease selection. American Journal of Physical Anthropology, **65**: 201-212.

Harpending, H. C. (1974). Genetic structure of small populations. Annual Review of Anthropology, **3**, 229-243.

Harpending, H. and L. Wansnider (1982). Population structure of the Dobe !Kung, In: M. Crawford & J. Mielke (eds.), Current Developments in Anthropological Genetics, Vol. 2, pp. 29-50. New York: Plenum.

Harpending, H. C. and Ward, R. H. (1982). Chemical systematics and human populations, In: M. H. Nitecki (ed.), Biochemical Aspects of Evolutionary Biology, pp. 213-256. Chicago: University of Chicago Press.

Harrison, G.A. and Boyce, A.J. (eds.) (1972). The Structure of Human Populations. Oxford: Clarendon.

Hiorns, R.W., Harrison, G.A. and Gibson, J. B. (1977). Genetic variation in some Oxfordshire Villages. Annals of Human Biology, **4**: 197-210.

Jorde, L. B. (1980). The genetic structure of subdivided human populations, In: J. Mielke & M. Crawford (eds.), Current Developments in Anthropological Genetics, Vol. 1, pp. 135-208. New York: Plenum.

Jorde, L. B. (1984). A comparison of parent-offspring and marital migration data as measures of gene flow, In: A. J. Boyce (ed.), Migration and Mobility: Biosocial Aspects of Human Movement, pp. 83-96. London: Taylor & Francis.

Jorde, L. B. (1985). Human genetic distance studies: present status and future prospects. Annual Review of Anthropology, **15**.

Jorde, L. B., Pitkanen, K. & Mielke, J. H. (1986). Migration in Kitee, Finland: genetic and epidemiologic consequences. Paper presented at the American Association of Physical Anthropology Meetings, Albuquerque, New Mexico, April 1986.

Lasker, G. W. (1985). Surnames and Genetic Structure. Cambridge: Cambridge University Press, 148 pp.

Leslie, P. M. (1985). Potential mates analysis and the study of human population structure. Yearbook of Physical Anthropolgy, **28**, 53-78.

Mielke, J. H. and Crawford, M. H. (eds.) (1980). Current Developments in Anthropological Genetics, Vol. 1: Theory and Methods. New York: Plenum. 436 pp.

Mielke, J. H., Jorde, L. B., Trapp, P. G., Anderton, D. L., Pitkanen, K. and Eriksson, A. W. (1984). Historical epidemiology of smallpox in Åland, Finland, 1751-1899. Demography, **21**, 271-295.

Menozzi, P. and Cavalli-Sforza, L. L. (1978). Synthetic maps of human gene frequencies in Europeans. Science **201**, 786-792.

Morton, N. E. (1969). Human population structure. Annual Review of Genetics, **3**, 53-73.

Lees, F. and Relethford, J. (1982).   Population structure and anthropometric variation in Ireland during the 1930's, In: M. H. Crawford & J. H. Mielke (eds.),   Current Developments in Anthropological Genetics,   Vol. 2, pp. 385-428. New York: Plenum.

Relethford, J. H. (1985).   Density-dependent migration and human population structure in historical Massachusetts.   American Journal of Physical Anthropology, **69**, 377-388.

Rogers, A. R. and Harpending, H. C. (n.d.) Bias in migration matrix models. (In press.)

Sattenspiel, L. (in press).   Population structure and the spread of disease. Human Biology.

Schull, W. J. and MacCluer, J. W. (1968).   Human genetics: structure of population. Annual Review of Genetics. Palo Alto: Annual Reviews Inc.

Sheldon G. (1895).   History of Deerfield (2 vols.).   Greenfield: Hall (republished 1972, New Hampshire Publishing Co.).

Smith, M. T. (1984).   The effect of migration on sampling in genetical surveys, In: A. J. Boyce (ed.), Migration and Mobility: Biosocial Aspects of Human Movement, pp. 97-109. London: Taylor & Francis.

Smouse, P. E., Long, J. C. and Sokal, R. R. (1987).   Multiple regression and correlation extensions of the Mantel Test of Matrix Correspondence.   (In press.)

Susanne, C. (1984).   Biological differences between migrants and non-migrants, In: A. J. Boyce (ed.),   Migration and Mobility: Biosocial Aspects of Human Movement, pp. 179-193. London: Taylor & Francis.

Swedlund, A. C. (1978).   Historical demography as population ecology.   Annual Review of Anthropology, **7**, 137-173.

Swedlund, A. C. (1980).   Historical demography: applications in anthropological genetics,   In: J. Mielke & M. Crawford (eds.), Current Developments in Anthropological Genetics, vol. 1, pp. 17-42. New York: Plenum.

Swedlund, A. C. (1984).   Historical studies of mobility, In: A. J. Boyce (ed.), Migration and Mobility: Biosocial Aspects of Human Movement, pp. 1-18. London: Taylor & Francis.

Swedlund, A. C., Jorde, L. B. and Mielke, J. H. (1984).   Population structure in the Connecticut Valley,   I: Marital migration.   American Journal of Physical Anthropology, **65**, 61-70.

Swedlund, A. C., Meindl, R. S. and Gradie, M. I. (1980).   Family reconstitution in the Connecticut Valley: Progress on record linkage and the mortality survey, In: B. Dyke & W. Morrill (eds.) Genealogical Demography, pp. 139-145. New York: Academic Press.

Swedlund, A. C. and Boyce, A. J. (1983).   Mating structure in historical populations: Estimation by analysis of surnames. Human Biology, **55**, 251-262.

Swedlund, A. C., Anderson, A. B. and Boyce, A. J. (1985).   Population structure in the Connecticut Valley, II: A comparison of multidimensional scaling solutions of migration matrices and isonymy.   American Journal of Physical Anthropology, **68**, 539-547.

Swingland, I. R. and Greenwood, P. J. (eds.) (1983).   The Ecology of Animal Movement. Oxford: Clarendon Press.

Wood, J. and Smouse, P. (1982).   A method of analyzing density-dependent vital
    rates with an application to the Gainj of Papua New Guinea.   American
    Journal of Physical Anthropology,   **58**,   403–411.

Wood, J., Smouse, P. and Long, J. (1985).   Sex-specific dispersal patterns in two
    human populations of Highland New Guinea.   American Naturalist, **125**,
    747–768.

MARRIAGE DISTANCE AND ETHNICITY

C. PEACH[1] and J. CLYDE MITCHELL[2]

[1] *St. Catherine's College, Oxford,*
[2] *Nuffield College, Oxford*

INTRODUCTION

This paper deals with the effect of residential segregation and marriage distance on social interaction. In the first part the significance of distance as a factor in marriage patterns is examined. In the second, multivariate analysis compares the contribution of ethnicity, education and distance to the observed patterns of marriage.

Western urban societies show three marked characteristics: (a) social and ethnic homogamy; (b) short marriage distances, and (c) social and ethnic residential segregation. There is some controversy in the literature as to how, if at all, these factors relate to one another (Ramsoy, 1966; Peach, 1974). It is argued here that, since social interactions, such as marriage, show strong distance decay effects, it should be possible to predict patterns of ethnic outmarriage from a knowledge of the spatial separation of ethnic groups. This hypothesis is tested for San Francisco using 1980 marriage certificate and census data.

DISTANCE DECAY

Studies of the marrying population in western urban societies universally indicate large numbers of marriages taking place over short distances (measured between the residences of the brides and grooms at the time of marriage) and decreasing numbers of marriages taking place with increasing distance. Such studies have been published by Bossard (1932) on Philadelphia, by Kennedy (1943) on New Haven, by Koller (1948) on Cleveland, Ohio, by Catton and Smircich (1964) on Seattle, by Ramsoy (1966) on Oslo, Norway, and by Coleman (1977, 1979) on British data.

There is a substantial literature demonstrating that not only is the *amount* of social interaction much greater, over short distances than random processes would suggest, but that the *degree* of such interaction is also much greater. If populations were uniformly distributed, the potential number of partners for interaction would increase geometrically with the square of the distance from a point.   Population is not, of course, uniformly distributed within cities. Nevertheless, on the local scale, the work of Festinger, Schachter and Back (1950) on new MIT graduate student apartment housing in Boston, demonstrated that the degree of selection of friends out of the available population was twice as high over their first unit of distance as over the second, and four times as high over the first as over the fourth.   Thus not only is the amount of interaction greater over short than long distances, but the degree to which those living close are selected is also greater.

Catton and Smircich's (1964) study was the first to demonstrate a similar effect in marriage distances.   Not only was there a strong distance decay effect but there was substantial overselection, statistically speaking, of the available partners within the first three miles of the residence of marrying population and underselection of the numbers available at increasing distance. Ramsoy's work on Oslo data produced very similar results (Table 1), although the overselection was confined to the even shorter distance of two miles.

Table 1   Observed and expected numbers of couples according to distance between residences before marriage: Oslo, 1962.

Distance (miles)	Observed marriages	Expected marriages	Ratio of observed to expected
<0.50	267	114.2	2.34
0.50-0.99	427	203.0	2.10
1.00-1.49	339	291.1	1.16
1.50-1.99	319	305.9	1.04
2.00-2.49	239	285.2	0.84
2.50-2.99	220	279.1	0.79
3.00-3.49	219	285.7	0.77
3.50-3.99	153	229.4	0.67
4.00-4.49	119	137.4	0.87
4.50-4.99	83	141.9	0.59
5.00-5.49	54	93.0	0.58
5.50-5.99	46	60.2	0.76
6.00-6.49	33	51.0	0.65
6.50-6.99	23	38.3	0.60
7.00-7.49	17	24.7	0.69
7.50-7.99	6	15.7	0.38
>8.00	9	17.2	0.52
Total:	2573	2573.0	

Source: Ramsoy, 1966.

It seems, therefore, reasonable to hypothesise that if people interact more strongly, on the whole, with those that are close to them rather than with those who are at some distance and, if groups are segregated according to social class or ethnicity, then (1) segregated groups would manifest high degrees of inmarriage because those living close would also be like them, and (2) dispersed groups would marry out among those groups among whom they were dispersed. There is some evidence from Australian studies that the expected relationship of segregation and inmarriage exists (Timms, 1969; Peach, 1974).

Duncan and Lieberson (1959) demonstrated for Chicago that there seemed to be a higher degree of inmarriage for those living in more segregated conditions and a higher degree of outmarriage for the more dispersed groups. Their data were not ideal. Parentage was used instead of marriage data. Their data demonstrated that spatially dispersed groups had a higher percentage of persons with one native born parent than the more segregated groups. Their assumption was that a higher proportion of native born parents represented more marriage into the host population, although, as they admitted, it was quite possible for the native born to be second generations of the same ethnic stock as the foreign born. They were unable to produce evidence of which particular groups were intermarrying.

Duncan and Lieberson's findings had important implications. If ethnic groups did interact with those groups among which they were residentially dispersed then one of the established theories of American assimilation, the triple melting pot (Kennedy, 1944, 1952) seemed improbable. According to this, it was argued that European ethnic groups would lose their national identity but maintain their religious identity. Thus the British, Germans and Scandinavians were supposed to merge within a Protestant melting pot; the Irish, Poles and Italians were supposed to merge within a Catholic pot, while eastern and western European Jews were supposed to form a Jewish melting pot. However, evidence of residential segregation in American cities provided consistent evidence (Lieberson, 1963) that the Irish were residentially mixed with the British, Germans and Scandinavians and not with the Italians and Poles, while Italians and Poles showed fairly marked degrees of residential segregation from each other.

Peach (1980) used the findings of the Duncan and Lieberson study to show that ethnic groups did marry into those among whom they were residentially mixed. He correlated matrices of ethnic intermarriage in New Haven with matrices of ethnic segregation for the same pairs of groups, concluding that geography operated as a key independent variable. Peach, however, measured segregation rather than physical distance between marriage partners.

The problem remained, therefore. Did proximity condition the marriage of like persons so that marrying near meant marrying like; or was the likeness of prospective spouses reflected in residential proximity so that marrying like meant also marrying near? Was geography an independent or a dependent variable? Were the two phenomena of marrying near and marrying like due to covariance or to causal linkage?

The most imaginative attempt to separate the factors of homogamy and propinquity was made by Ramsoy (1966) who claimed that the two tendencies were totally independent of one another. Ramsoy demonstrated from Oslo data that (1) marriage between occupationally similar spouses occurred more often than random expectation; (2) that spouses were statistically overselected from those who were residentially proximate; (3) that the population of Oslo was residentially segregated by occupational class. She then claimed that segregation was not the structure which held like marriages and near marriages together. Her analysis showed that the same proportion of like marriages took place irrespective of the distance apart of the partners; and also that the median distance of marriage remained constant independent of the degree of social similarity of the partners. She claimed that the tendency to marry like and the tendency to marry close were independent.

Peach (1974) suggested a flaw in Ramsoy's argument. He pointed out that Ramsoy's system of classifying degrees of likeness or unlikeness in marriage might eliminate the effects which she was seeking to measure. Ramsoy's categorisation of like marriage meant, for example, that doctors marrying doctors and factory workers marrying factory workers would be included in a single category of like marrying like. Since Koller's (1948) evidence had shown that higher occupational groups had longer marriage distances than lower occupational groups, Peach argued that placing long and short distance groups into a single category would produce an average marriage distance of all the groups concerned. As the degree of social unlikeness increased, mean marriage distances would nevertheless remain constant. (The marriage distance of two people marrying outside their respective classes might be supposed to be the mean of the sum of the in-marrying distance for each class). The mean distance of the combination of a long class and a short class would be an intermediate distance.) Similarly, although the proportion of like marriages might remain constant irrespective of distance, the class composition of like marriages might vary considerably. Thus a higher proportion of like marriages at low distances could be composed of lower classes and a higher proportion of marriages at longer distances could be composed of higher classes.

To summarise, there is a substantial literature demonstrating the overselection of socially similar partners in marriage. There is a considerable literature demonstrating the overselection of spouses from geographically proximate areas; there is a considerable literature demonstrating the spatial segregation of society on social and ethnic groups. There is some evidence that spatial segregation correlates with ethnic intermarriage, but there is no clear evidence of whether the connexion is causal or whether both the ethnic patterning and the ethnic marital choices stem from common attitudes rather than being independently interlinked.

THE SAN FRANCISCO DATA

An attempt was made to resolve some of these problems through an analysis of San Francisco marriage certificates for the year 1980. The certificates for 1980 were chosen so that information could be cross checked with the census data for the same year. San Francisco was chosen as the location of the study for three reasons: (a) the city is of a manageable size, 700,000 persons; (b) the city contains a wide range of ethnic groups, Whites (58 per cent), Hispanics (12 per cent), Asians (22 per cent) and Blacks (13 per cent); (c) California is thought to have much more relaxed attitudes to social experimentation than other parts of the United States so that it was thought that trends in intermarriage which might be established there might spread to other parts of the country at later times.

There were in 1980 in San Francisco 6,578 marriages. Complete counts of marriage certificates for the months of January, April, July and October 1980 yielded 2,104 records (32 per cent of the total number of marriages recorded for the city). The marriages sampled contained 399 cases where neither party resided in San Francisco or its immediate environs. These records were eliminated from consideration since the exercise was designed to examine the effect of residential location in San Francisco on marriage. This left a total of 1,705 usable records.

Information on the certificates was coded for the ethnicity of the groom's father and mother, the bride's father and mother, groom's generation (first, second or third), bride's generation; groom's residential location, bride's residential location; the distance between the groom and bride's location, the number of years of schooling recorded for the groom and for the bride.

Ethnic identification was made from a number of different sources of information contained in the certificates. Just under 50 per cent of the brides and grooms were first or second generation so that information about parental or

Table 2:   Ethnic intermarriage, San Francisco 1980: brides' ethnicity by columns; grooms' ethnicity by rows

(groom ↓ / bride →)	Anglo	German	Irish	Italian	Other European	Mexican	Other Latin American	Chinese	Japanese	Filipino	Other Asian	Middle East	Jewish	Black	Other	Total
Anglo	186	39	23	22	35	15	14	17	6	11	4	2	25	1	2	402
German	45	28	11	9	11	7	8	5	2	5	3	1	3	0	1	139
Irish	44	10	13	5	4	1	2	1	2	1	1	1	4	0	0	89
Italian	22	10	7	17	8	2	4	1	1	0	0	0	4	1	1	78
Other European	35	12	9	6	40	1	4	0	2	0	0	1	6	0	0	116
Mexican	16	4	7	5	4	52	17	1	1	0	0	0	2	3	3	120
Other Latin American	16	5	2	4	4	7	78	1	0	6	0	0	5	0	0	119
Chinese	3	0	2	1	0	2	1	143	4	0	4	0	0	0	1	165
Japanese	7	1	1	0	0	1	0	9	14	1	3	1	0	0	1	39
Filipino	6	2	0	4	0	2	3	3	1	75	0	0	1	1	1	98
Other Asian	2	2	2	1	1	1	3	3	1	1	26	0	0	0	1	44
Middle East	11	0	4	1	4	1	3	0	0	0	0	22	1	0	0	50
Jewish	27	4	5	6	0	5	2	2	2	1	0	0	55	0	3	119
Black	2	1	0	2	1	4	3	0	0	2	2	0	1	97	1	116
Other	0	0	0	0	1	0	2	2	0	1	0	0	0	0	5	11
Total:	422	118	86	81	123	101	144	188	36	104	43	28	110	106	15	1705

personal birthplace was a major identifying source. Family names and personal names were the most important element in identification and there was supplementary information from the type of marriage ceremony in some cases. Altogether, 69 ethnic categories were recorded. However, for the sake of analysis, they were reduced to 15 major categories in which ethnic identity of each individual was defined as that of the father (Table 2).

Totals for each ethnic group were compared with the census total insofar as comparable groups were covered (Jews, for example, do not appear as a category in the US census).

On this basis it appeared that while there was general congruence, the Anglo group was larger than expected in the marriage certificates than in the census, while the black population was undercounted. Given the difficulty of identifying the black population from marriage certificate information, it is probable that many were classified as Anglo and possible also that some Anglos were classified as black. (While this problem affects the estimates of the ethnic totals, it seems unlikely to have a great impact on the measure of marriage between groups). The Bureau of the Census estimated that of the total number of 49,714 marriages in the USA in March 1980 only 167 involved Black/White couples (US Bureau of the Census, 1982, 42).) Addresses were coded for brides and grooms before the detailed ethnic geography of San Francisco had been investigated, but it was clear that the large majority of those who had been identified from marriage certificate information as black did in fact reside in areas shown in the census to be areas of dense black settlement. Similarly, indices of residential dissimilarity calculated for comparable groups from census and marriage certificate data correlated well with each other. Thus, although the classification is not perfect, comparisons of independently collected information suggest that for most groups there is a good degree of agreement. The comparison of census and marriage certificate ethnic identity is given in Table 3.

Apart from the estimation of ethnicity, there are several other data problems which require review. Many studies confine themselves to marriages where both partners reside in the city and exclude marriages which take place across the boundaries with suburbs. If one were to do this with the San Francisco data, the number of marriages would decrease from 1,705 to 1,220.

A second problem is that of same-address couples. A small sample of San Francisco marriages in 1950 indicated an occurrence rate of 15 per cent. Catton and Smircich (1964) found that 11 per cent of their usable sample of marriages in

Table 3:   Comparison of census and marriage certificate data for selected
               ethnicities, San Francisco, 1980.

Ethnic group	Census No.	Census %	Grooms No.	Grooms %	Brides No.	Brides %
Total	678924	100.00	1705	100.00	1705	100.00
Asian	163808	21.71	346	20.29	371	21.76
Black	86414	12.73	112	6.57	104	6.10
European (SA & MA)	362782	53.43	824	48.33	830	48.68
Spanish Origin	83373	12.28	239	14.02	245	14.37
Japanese	12046	1.77	39	2.29	36	2.11
Chinese	82480	12.15	165	9.68	188	11.03
Filipino	38265	5.63	98	5.75	104	6.10
Korean	3763	0.55	10	0.59	19	1.11
Vietnamese	5583	0.82	14	0.82	15	0.88
Mexican	32633	4.81	120	7.03	101	5.92
Puerto Rican	5174	0.76	7	0.41	14	0.82
Cuban	1397	0.20	6	0.35	4	0.23
Other Latin	44169	6.51	106	6.22	126	7.39
English (SA)	33703	4.96	402	23.58	422	24.75
English (MA)	52623	7.75				
Scottish (SA)	3409	0.50				
French (SA)	6996	1.03	28	1.64	32	1.88
French (MA)	20113	2.96				
German (SA)	25673	3.78	139	8.15	118	6.92
German (MA)	48802	7.19				
Irish (SA)	31248	4.60	89	5.22	86	5.04
Irish (MA)	55290	8.14				
Italian (SA)	27424	4.04	79	4.63	82	4.81
Italian (MA)	14451	2.13				

Note: SA = Single Ancestry;   MA = Multiple Ancestry.
       The multiple ancestry group is not exclusively
       counted. This is to say that someone of Anglo-
       German ancestry, for example, may be included in
       both the English and German multiple ancestry group.

Seattle were living at the same address.   Ramsoy (1966) found that about 17 per
cent of her 1962 Oslo sample shared the same address.   Although the
calculations of base populations on which these percentages were calculated
differ, one could say that the proportion of same address marriages probably lay
between 10 and 20 per cent of the total in the 1950s and 1960s. However, in San
Francisco by 1980, 55 per cent of couples gave the same address on their
marriage certificates.

Table 4 gives the distances in mile intervals for the 1705 San Francisco marriages. There is a sharp decline from 153 in the first mile to less than half of this figure between 2.01 and 2.50 miles. Thereafter, it rises to 73 marriages between 3 and 4 miles then drops to a plateau of 50 until after 6 miles, when there is a further sharp decline (Table 4).

MULTIVARIATE ANALYSIS

From the 15*15 matrix of inter-ethnic marriage frequencies for 1980, a multidimensional analysis of the dissimilarity indices enabled us to represent the separation of the ethnic groups in terms of the extent to which the marriage patterns between ethnic groups were similar. The pattern may be conveniently represented by a dendrogram derived from the coordinates determined by the multi-dimensional scaling (Figure 1).

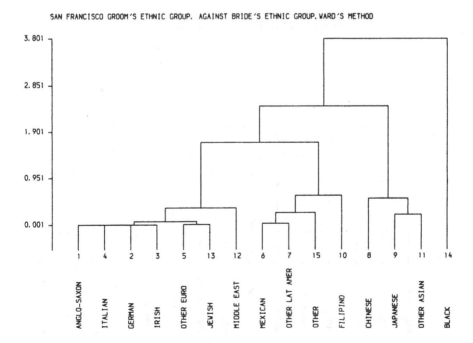

Figure 1: San Francisco Groom's ethnic group, against Bride's ethnic group, Ward's method

The extent to which different background factors may have influenced the frequency of inter-ethnic marriage, however, requires the basic 15*15 table of inter-marriage frequencies to be cross-tabulated against these background factors. In order to simplify analysis we have reduced the three basic structural factors available to us in the data, to expression of 'distance'. For example, in respect of the influence of ethnicity *per se* we generated a dendrogram on *a priori* grounds so as to reflect the social and cultural separation of the 15 broad ethnic groups we have decided to work with. Figure 2 is a representation of our judgement of the social and cultural distance separating these ethnic groups.

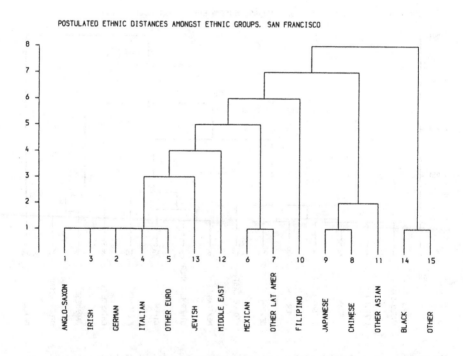

POSTULATED ETHNIC DISTANCES AMONGST ETHNIC GROUPS. SAN FRANCISCO

Figure 2:   Postulated ethnic distances amongst ethnic groups, San Francisco.

The distance separating any two groups may now be reflected by the 'height' of the node at which the two groups are linked together. At the first level each group is linked to its own group by the smallest distance, here arbitrarily given a distance of 1. The main European groups, however, would be linked at the next distance, i.e. at 2. The group reflected as Jewish would link in with the General European cluster at 3. At the other extreme we estimated that the Blacks and Others would link in with the overall European group at the maximum arbitrary distance of 8. It should be noted that these 'distances' are not deemed to have any *cardinal* value: they merely represent categorical labels which may be used as a third dimension in a cross-tabulation in which the ethnic inter-marriage frequencies are controlled for by social and cultural distances which we assume separate the groups. A test of the validity of our *a priori* assumptions will emerge from the subsequent analysis.

The analytical procedure we have adopted is to estimate the fit of cell frequencies in the 15*15*8 table using log-linear methods (see Gilbert, 1981). The base line chi-square may be taken as that derived from the assumption that each cell of the table is filled with a constant value which sums to the total of 1705. This is the extreme 'no fit' value of the chi-square. In respect of ethnic distance this value is 11709 with 1799 degrees of freedom. The effect of the frequencies of grooms and brides of the appropriate ethnic groups on the frequencies of the cells can now be estimated in the table. By fitting the marginals for grooms and brides the chi-square drops from the estimated 11,709 to 9982 with 1771 degrees of freedom, i.e. a loss of 28 degrees of freedom which is one less than the total number of 15 groups for grooms and one less for the total number of 15 groups for brides. The significant statistical improvement of 1827 in chi-square for 28 degrees of freedom merely chronicles what we would expect on commonsense grounds, viz. that the total number of grooms and brides from different ethnic groups is going to influence the frequencies in each cell. When we take into account the effect of ethnic distance the chi-square drops from 9982 to 7881 with 1764 degrees of freedom. A fit as poor as this would not be acceptable to statisticians as an adequate modelling of the effects influencing the frequency of inter-ethnic marriage. Nevertheless we may observe that an improvement in chi-square of 2101 for a loss of 7 degrees of freedom suggests that ethnic distance as we have defined it is having an appreciable effect on the frequency of inter-ethnic marriage. The decrease in chi-square of 2101 over the chi-square when only the marginals of the number of grooms and brides of different ethnic groups is taken into account represents an improvement of 21%.

A test of the validity of our estimate of what we have called ethnic distance is provided by an examination of the coefficients estimated from our model for the distance categories derived from the dendrogram in Figure 2. These are in order, category i. = 1.0; category ii. = -0.716; category iii. = -2.203; category iv. = -3.414; category v. = -1.826; category vi. = -2.962; category vii. = -2.172; and category viii. = -3.109. These values are logarithmic estimates of the modifications to be applied to the cell frequencies to take account of the effect of the appropriate ethnic distance category. The negative values, therefore, represent an expected diminution in the cell frequency associated with that ethnic distance category. The first three values are pleasingly monotonic, but the value for category v is slightly out of line, for category vi, vii still out of line. The distance v is associated with the point at which we judged that the Mexicans and Other Latin Americans are linked to the people of European parentage, the Jewish people and the people from the Middle East. Clearly we have misjudged the extent to which the Mexicans and Other Latin Americans are socially and culturally separated from the other ethnic groups. Re-ordering the groups in accordance with the coefficients would be: i. All groups inter-marrying within their own groups. ii. Groups marrying within sub-clusters, e.g. the European groups amongst themselves. iii. The Mexicans and Other Latin Americans inter-marrying with the major European cluster. iv. The Japanese, Chinese and Other Asian groups intermarrying with the Europeans and Latin Americans. v. The Jewish people inter-marrying with the Europeans, Latin Americans and Jewish people. vi. The Filipinos inter-marrying with the peoples already listed. vii. The Blacks and others inter-marrying with those already listed, and finally viii. The people of Middle Eastern extraction inter-marrying with all the other groups - possibly due to Islam.

We may use the same basic procedure to estimate the effect of geographic distance on the probability of intermarriage amongst ethnic groups. The geographic distances separating the residences of grooms and brides before marriage were estimated directly from the geographic locations of their addresses on a map. These were divided into nine categories (as in Table 4, with marriages over six miles grouped together, and with the addition of marriages anywhere outside San Francisco).

As before, we may cross-tabulate the frequency of ethnic inter-marriages by the categories of distances separating the places of residence of the spouses. The effect of the frequencies of grooms and brides on the cell frequencies was taken into account, before examining the part played by residence distance on inter-marriage frequency. In this case the improvement of chi-square is from 5911.3 with 1996 degrees of freedom to 3957.4 with 1988 degrees of freedom representing an improvement of chi-square of 1953.9 for 8 degrees of freedom.

This represents an improvement of 33.1% in chi-square, which is proportionately greater than the effect of ethnic distance with the same cross-tabulation of grooms' and brides' frequencies. The coefficients reflecting the effect of geographic separation of spouses are in order: i. Living together = 1.000. ii. 1 mile or less = -1.759. iii. 1 to 2 miles = -2.204. iv. 2 to 3 miles = -2.599. v. 3 to 4 miles = -2.499. vi. 4 to 5 miles = -2.819. vii. 5 to 6 miles = -2.838. viii. 6 miles or more = -1.360. ix. Outside San Francisco = -2.246. The effect of distance on the frequency of inter-marriage seems to tail off beyond a distance of 5 miles. Otherwise the effect seems reasonably consistent and monotonic.

Finally, the effect of the difference in educational background on the frequency of ethnic inter-marriage was assessed using the same techniques. The educational levels of grooms and brides were available from the records. The separation of educational level of brides and their grooms was classified into nine categories: i. educational standard of both spouses at high school level; ii. both at college level; iii. both at post-college level; iv. groom at high school, bride at college level; v. groom at college, bride at post-college level; vi. groom at high school, bride at post-college level; vi. groom at college, bride at post-college level; vi. groom at high school, bride at post-college level; vii. groom at college, bride at high school level; viii. groom at post-college, bride at high school level; and, finally, ix. groom at post-college, bride at college level.

The logic of this arrangement is that the extreme off-diagonal positions of vi and ix represent the greatest separation of educational level, the intermediate positions of iv, v, vii and viii less extreme educational separation, while i, ii and iii on the diagonal, similar educational standards at different levels. The positions iv, v and vi represent situations where the bride had higher educational qualifications than the groom, while positions vii, viii and ix those situations where the groom had higher educational qualifications than the bride.

The improvement of chi-square due to the effect of educational differences on the cell frequencies of inter-ethnic marriages was from 5368.8 with 1996 degrees of freedom to 4253.5 with 1988 degrees of freedom representing an improvement of 20.8%, which is similar to the effect of ethnic difference (21.0%) but slightly lower than the effect of geographical distance (33.1%). Of particular interest are the effects of the different degrees of educational difference on the cell frequencies. The following table sets out the logarithmic coefficients affecting the inter-marriage cell frequencies.

Groom's educational level	Bride's educational level		
	High School	College	Post-College
High school	1.000	-1.241	-2.986
College	-0.948	0.133	-1.683
Post-college	-2.364	-0.805	-1.165

The educational difference which has the largest effect on the frequency of inter-ethnic marriage is that when the marriage is between a high school groom and a post-college educated bride. The large negative logarithmic effect implies that the cell frequency of such marriages is reduced most of all when those circumstances prevail. Of almost the same degree of effect, however, is when a groom with post-college education marries a bride with only high school education. The effect on inter-ethnic marriages is relatively large when grooms with high school education marry brides with college education or alternatively when grooms with college education marry brides with post-college education. The effect in the converse situation, i.e. when brides are marrying grooms with relatively less education, is somewhat smaller. Lastly, as compared with the frequency of ethnic inter-marriages of spouses both with post-college education is likely to be even smaller.

In summary, the effects of three background factors which may affect ethnic inter-marriage frequencies have been tested. The greatest influence seems to be simply how far apart the spouses lived before they married. The effect of educational differences on the frequency of ethnic inter-marriages though appreciable was somewhat less. Lastly, somewhat surprisingly, the social and cultural differences separating the spouses, as estimated *a priori* from our understanding of the situation had the least effect of all three factors.

ACKNOWLEDGMENT

Acknowledgment is made to the ESRC for funding the field work for this research.

REFERENCES

Bossard, J.H.S. (1932).   Residential propinquity as a factor in marriage selection. American Journal of Sociology, **38**, 219-224.

Catton, W.R. & Smircich, R.J. (1964).   A comparison of mathematical models for the effect of residential proximity on mate selection.   American Sociological Review, **29**, 522-529.

Coleman, D. (1977). The geography of marriage in Britain, 1920-1960. Annals of Human Biology, **4**(2), 101-132.

Coleman, D. (1979).   A study of the spatial aspects of partner choice from a human biological view point. Man (NS), **14**, 414-435.

Duncan, O.D. & Lieberson, S. (1959). Ethnic segregation and assimilation. American Journal of Sociology, **64**, 64-374.

Festinger, L.,. Schachter, S. & Back, K. (1950). Social Pressures in Informal Groups. New York.

Gilbert, G.N. (1981). Modelling society: an introduction to loglinear analysis for social researchers. London: George Allen & Unwin.

Kennedy, R.J.R. (1943). Premarital residential propinquity and ethnic endogamy. American Journal of Sociobiology, **48**, 580-584.

Kennedy, R.J.R. (1944). Single or triple melting pot? Intermarriage in New Haven, 1870-1940. American Journal of Sociology, **49**, 331-339.

Kennedy, R.J.R. (1952). Single or triple melting pot? Intermarriage in New Haven, 1870-1950. American Journal of Sociology, **58**, 56-59.

Koller, M.R. (1948). Residential proximity of white mates at marriage in relation to age and occupation of males, Columbus, Ohio, 1938 and 1946. American Sociological Review, **13**, 613-616.

Lieberson (1963). Ethnic Patterns in American Cities. Glencoe: The Free Press.

Peach, C. (1974). Homogamy, propinquity and segregation: a re-evaluation. American Sociological Review, **39**, 636-641.

Peach, C. (1980). Ethnic segregation and intermarriage. Annals of the Association of American Geographers, **70**, 371-381.

Ramsoy, N.R. (1966). Assortative mating and the structure of cities. American Sociological Review, **31**, 773-786.

Timms, D.W.G. (1969). The dissimilarity between overseas-born and Australian-born in Queensland: dimensions of assimilation. Sociology and Social Research, **53**, 363-374.

U.S. Bureau of the Census (1962). Statistical Abstract of the United States: 1982-83, Washington D.C.

MATING PATTERNS IN ISOLATES

L. JAKOBI and P. DARLU

Centre de Recherches Anthropologiques, Musée de l'Homme, Paris, and
Unité de Recherches de Génétique Epidémiologique, INSERM, Paris.

INTRODUCTION

Various factors play a role in the biological evolution of a human group, such as historic events, migration, choice of spouse, selection and random transmission of genes. All these factors are elements of a complex and interactive whole where demographic factors, social relations, economic exchanges and cultural behaviours influence one another.

However, in this complex whole, marriage has a special role, since it guarantees biological, cultural, and socioeconomic continuity. That is why the study of mating patterns has excited so much interest in different disciplines. For biological anthropologists and population geneticists, isolates represent a special situation in which to understand the problem of mate choice and its implications and consequences.

Two important questions can be asked with regard to the setting up and the effectiveness of the mechanisms which influence the formation of unions. (1) What are the constraints which act on the choice of the spouse and what are the limitations of this choice? (2) What are the possible consequences of the choice of mate and of matrimonial migration on the diffusion of genes and the evolution of the gene pool?

RESTRICTIONS IN THE CHOICE OF A SPOUSE

In a small population, various constraints - demographic, social, cultural or geographic - determine the choice of a spouse. As a first observation, the choice of a spouse is necessarily limited in an isolate, on account of *demographic factors*. The probability of marrying a relative is not negligible, even if marriages are undertaken at "random"; for example, it can be demonstrated that the frequency p_d of marriage between cousins of degree d may be written (Dahlberg, 1948; Malecot, 1948; Jacquard, 1970):

$$p_d = \frac{2^{2d+1}}{N_g} \left[1 - \frac{1}{m_g} + \frac{\sigma_g^2}{m_g^2} \right]$$

where N, m and σ^2 are respectively the population size, the average number of children and its variance, at the generation g from the common ancestors of the first cousins (d=1) or of the second cousins (d=2).

This frequency is inversely proportional to the size of the population, i.e. when the size is small the frequency is large. It is also a function of the mean and variance of the number of children per family. It is therefore directly related to the demographic fluctuations of the population (expansion or extinction), but also to the diversity of conjugal behaviour with respect to the number of children (Table 1).

From a genetic point of view, this equation does not give the exact proportion of consanguineous marriages defined by the value of the kinship coefficient between the two spouses. As a matter of fact, in an isolate, this coefficient increases with the number g of past generations since the isolation of the population, and decreases with the population size. The expected kinship coefficient (1-exp(g/2N)) converges to 1, and its variance ((1/6N)exp(-2g/N)) converges to zero as g increases (Jacquard, 1975). The more complete the genealogical information over several generations, the more the kinship coefficient between spouses tends towards identity, whether they are just cousins or more distant relatives. The probability that a man will marry a woman who is neither a first cousin nor a second cousin, but nonetheless has a high kinship coefficient with her, eventually higher than with a real cousin, is therefore not negligible *a priori* if the population is limited in size and has been isolated for a long time.

Table 1:   Number of cousin marriages in an isolate population (N = 1000) according to different demographic hypotheses (French village of Arthez d'Asson).

	Dahlberg formula ($\sigma_g^2 = 0$)	Poisson distribution ($m_g = \sigma_g^2$)	Observed distribution ($m_g \neq \sigma_g^2$)
First cousins	6	8	10
Second cousins	23	32	40
Third cousins	88	128	158

m_g and σ_g^2 are respectively mean and variance of the progeny number.

The *age difference between spouses,* which is the general rule in many societies, makes up an additional demographic constraint which offers a man more possibilities for union with matrilateral cross cousins than with patrilateral cross cousins (Hajnal, 1963; Barrai et al, 1962). The extent of this phenomenon depends on the average number of different types of relatives for a given individual and of their age distributions, on the variance of number of children per family, on the distribution of age differences between spouses, on the demographic dynamics of the population, and on the degree of endogamy.

In certain societies, what may appear to be a preferential union, culturally encouraged, may in fact be no more than a simple consequence of a particular demographic situation.   Cazes (1981), in her study of the Dogon of Tabi in Mali, demonstrates how the proportions of the different types of unions between cousins do not correspond to what strict tradition requires (Table 2). She shows also that the proportions of marriages between different types of cousins adjust easily to the theoretical proportions given by Hajnal, i.e. under random marriage considering the age difference between spouses.   Comparing all the marriages that are theoretically possible with the marriage which is actually contracted, Cazes observes that only the daughter of the mother's sister seems to be left out, while all the other marriages appear to be represented without significant bias.

Another factor constraining the choice of a spouse is the *geographical distance* which separates the birth places and the places of residence of each of the spouses.   It is assumed that the probability of choosing one's partner diminishes when this distance increases, according to a function (exponential, natural log, gamma, etc.) which is specific for each situation.

Table 2: Comparison between observed and expected frequencies under a panmictic model of various types of marriage between close relatives or between people classified as cousins.

	Paternal cross cousins*	Paternal parallel cousins	Maternal parallel cousins	Maternal cross cousins**
Observed frequency	.112	.235	.008	.192
Frequency under panmictic model	.134	.213	.059	.152

* and ** are the preferred unions as expressed in the oral tradition of the group, respectively today and formerly.   Dogon populations of Tupéré (from Cazes, 1981).

The distribution of distances between birthplaces of the two spouses has been related to the level of neighbourhood knowledge and the degree of contact with their social surroundings (Boyce et al, 1967;   Cavalli-Sforza, 1963). Examples of these relationships have been supplied by Fix (1974) with respect to the Semai of Malaysia, and by Cavalli-Sforza and Hewlett (1982) for the Aka pygmies of Central Africa.

In their studies of Peul Bande, Pison and Lathrop (1982) analysed the effect of distance on the choice of a partner.   In this traditional society, the man's choice is preferential: he usually selects a wife who is related to him in patrilineal fashion (lineage endogamy) and furthermore practises village endogamy.   As a matter of fact, the two authors show that one can explain the lineage endogamy as a simple consequence of the smallest geographic distance between individuals belonging to the same lineage.

Sutter (1958) demonstrated that the proportion of spouses having their origins in the same village is only slightly higher when the marriage is consanguineous (39.5%) than when the marriage is not consanguineous (36.4%) (statistics obtained from a sample of 546 consanguineous marriages between 1919 and 1925 in a French department).   In addition, consanguineous marriages are more numerous when the spouses originate from two different villages of the same department (42.4% instead of 39.5%).   Besides, those who reside in a village where marriages between cousins have been noted find their spouse at greater distances than do those who live in villages without consanguinity: the most endogamous villages are also those where the matrimonial migration extends the furthest in geographical terms.

The choice of a mate also can be based on economic or cultural criteria. In Bearn (French Western Pyrenees), for example, it was customary until 1791, and certainly after, for the eldest child, male or female, to be the sole heir. With the aim of keeping the inheritance intact, the heirs had to marry preferentially someone who is not first-born.

Evidently this marriage rule is closely connected with the demographic variables. If the average number of children is very high, then the possibility of the numerous younger siblings finding a spouse who is a first-born is extremely small.   Hence celibacy and migration are inevitable.   The comparison between the total number and the effective number of children by family illustrates these phenomena in the case of Arthez d'Asson, a village in the Bearn (Table 3).   The interesting question is nonetheless to verify whether such a rule is actually and explicitly followed, or if it can be explained as well by a random choice under specific demographic situations.

Table 3 Assortative mating for social rank before and after 1825 in Arthez d'Asson, France.

	Before 1825 Male				After 1825 Male		
	1	2	>2		1	2	>2
Female 1	4.02	*4.07	0.52	Female 1	2.03	*0.00	-1.09
2	*3.11	2.86	-1.88	2	*1.75	0.53	0.68
>2	3.96	-1.40	-7.11	>2	-0.24	-0.28	-2.07

<table>
<tr><td>N = 476</td><td>N = 449</td></tr>
<tr><td>$\chi^2_1 = 122.5$; m = 3.79</td><td>$\chi^2_1 = 13.6$; m = 2.92</td></tr>
<tr><td>$\chi^2_2 = 103.9$; $\hat{m}$= 2.10</td><td>$\chi^2_2 = 69.6$; $\hat{m}$= 1.98</td></tr>
</table>

Values are the standardised deviates from random mating using the observed distribution of the progeny number of mean m. The sum of their squares gives χ^2_1. χ^2_1 is the minimum Chi square under the hypothesis of a Poisson distribution of the number of children by family. Its estimated parameter is $\hat{m}$. N is the sample size of observed unions. * = class of unions said to be the most traditional and preferred.

Until 1825, it seemed that there was a certain rule. As a matter of fact, one notices a large surplus of marriages between first-borns, which is a departure from custom, or between first-born and non-first-born; but there is conspicuous lack of intermarriages between non-first-borns. This marital behaviour is substantially transformed after 1825 where only marriages between first-born seemed to be preferred. However, after 1825 there was an important drop in the birthrate which reduced the average number of children per family from 3.8 to 2.9. Thus, statistically, it follows that the proportion of marriages between first-born, or between first-born and not-first-born, is greater. If the custom is still followed in the opinion of the inhabitants, it is largely because of these demographic changes.

CHOICE OF SPOUSE AND EVOLUTION OF THE GENETIC POOL

One use of genealogical reconstruction is that it allows the paths through which the collective gene pool has been transmitted, from generation to generation, to be followed. To write the genealogical history of a group amounts in some way to describing its biological history. Nonetheless, one must take account of the fact that the history of gene transmission cannot be reconstructed with certainty, but only with a degree of probability, because of the mechanism of sexual reproduction. As described by Mendel's laws, every gene has one chance in two of being transmitted and this is why the genealogical information can only be interpreted in terms of probabilities. Besides, this

information becomes very complex as one goes back in time.   In the case of
closed isolates, due to the existence of remote consanguinity, one readily
encounters genealogies that are inextricably embedded in complex networks.

The concept of probability of origin of genes (or POG) offers a simple,
convenient, and desirable means of using and of summing up this information
which is both rich and complex.   It permits the genetic evolution of a small
population to be characterised in a more detailed fashion than does the analysis
of consanguinity, the two methods being complementary (Serre, 1984).

This concept had its origin in the study made by Roberts (1968a,b) of the
population of an island in the South Atlantic ocean, Tristan da Cunha.   Using
genealogical information which dates from the island's first arrivals, i.e. the
founders, Roberts characterised the gene pool of the present population by the
"contribution" of the various founders.   Taking up this idea, Jacquard developed
a theory and introduced the concept of probability of origin of genes (Chapman &
Jacquard, 1971; Jacquard, 1977).   Given an individual from a population whose
genealogies are known over a large enough number of generations, a gene of this
individual could have come from each parent with a probability of 1/2, 1/4 from
each grandparent, 1/8 from each great-grandparent, and so on.   Summing up the
coefficients computed by going back through the different paths leading to the
last known ancestors, one obtains parameters w_{ij} which represent "the
probability that a gene of an individual i has been transmitted to him by the
founder j", i.e. "the part of the genetic constitution of individual i which comes
from the founder j".   These parameters can themselves be added up for the whole
population or for a part of it, e.g. one generation.   It is thus possible to follow
the transformation of the gene pool of a population as a function of time.   One
must point out here that, by convention, founders are defined as ancestors with
unknown parents.   They can be the historic founders at the initial establishment
of a group, the ancestors representing the limit of the genealogical information,
or immigrants joining the group at any time during its history.   In its
application, this concept has the advantage of providing a description of the
genetic melting pot through time as a function of the social and geographical
mobility of spouses.   It offers also the chance to investigate the role of descent
in the transmission of genes.

The probabilities of origin of genes were calculated for the first time by
Jacquard in the case of an Indian isolate in Honduras, studied by the
anthropologist Chapman who provided the genealogical information for the
genetic study (Chapman & Jacquard, 1971; Jacquard, 1977).   In 1870, eight of
these Indians, called Jicaques, took refuge in the mountains to escape the

Table 4: The Jicaque Indian isolate (Honduras). Probabilities of origin of the genes at each generation (per 1000). (From Jacquard, 1977.)

				Historic founders			
Generation	Juan	Polonia	Pedro	Petrona	Francisco	Faciana	Leon	Total
1	143	143	143	143	143	143	143	1000
2	171	171	229	229	100	100	0	1000
3	159	153	164	164	116	116	38	910
4	160	133	216	216	64	64	48	901
5	148	129	177	177	50	50	47	778
6	143	132	158	158	38	38	46	713

serfdom imposed on them by the Mestizos and the Spanish colonisers. At present, the group numbers about 300. The information gathered over a century concerns 1065 individuals.

In Table 4 which gives the POG (per 1000) at each generation, the founders represent generation 1, with an equivalent contribution: 143 per 1000. These founders are the three couples made up by Juan and Polonia, Pedro and Petrona, Francisco and Caciana (Polonia being the sister of Pedro, and Petrona being that of Francisco), and Leon, who subsequently married Juana, the daughter of Pedro and Petrona, and the eight persons of the initial group. Juana's name does not appear in the table. Having arrived as a child, she belongs to generation 2 as an adult. Her contribution and that of her husband Leon to this generation are thus equal to zero, their children being included in generation 3.

Two particular observations can be made. First, the disparity of contributions of the various founders. This is a general observation with respect to the different populations which have been studied. Such disparity is connected to the small number of children of some founders and also to the modest fertility of these children. Thus the share of Pedro in the genetic pool of the latest generation is 158 per 1000, while that of Francisco is only 38 per 1000. The second observation concerns the contribution of immigrant spouses (last column in Table 4). This reaches 287 per 1000 (287 representing the complement of 713 in 1000). Such a contribution is a very important and unexpected figure, given the hostility of the Jicaques to foreign intrusions. In fact, this hostility seemed very effective in reducing the entry of members from outside to the group: at each generation these represent only 4% of the married total. This influx, limited in number, is considered very modest by demographers and anthropologists, but also by the Jicaques themselves.

Nevertheless, as a consequence there was an important transformation of the genetic structure. This is due, principally, to the fact that couples who have one immigrant member have had higher fecundity, though this is not an iron-clad rule. Thus it is not the number of individuals who join the community from outside at each generation which is important, but the genes imported by these immigrants and diffused through their offspring in following generations.

By contrast, immigration of spouses was very high in the case of another isolate, an Eskimo isolate, located at Scoresby Sound on the east coast of Greenland. This isolate was founded in 1925 by 70 Eskimo Ammassalimiut who came from Angmassalik situated 1000 kilometers to the south. Subsequently, seven died just after their arrival in Scoresby Sound, which reduced the number of original founders to 63. Robert (1971) reconstructed the sociodemographic and genealogical history of this group whose number was 410 in 1970. The initial founders were identified and their parental relations known, thanks to the census and to the genealogies established previously in Angmassalik by Gessain (1970) in 1935-36 and in 1965. At Scoresby Sound, the genealogies concern 797 individuals. The total number of founders, original founders and immigrants who came later, is 124. Of them, 52 are Americans or Europeans (essentially Danish). The POG value (Jacquard, 1973, 1977) shows a great stablity in the contribution of original founders, i.e. the Ammassalimiut: 950 per 1000 in 1930, 892 per 1000 in 1970. The major part of the gene pool is derived from them, even if their share is different in terms of descent. But what is more remarkable is the very small "non-Eskimo" contribution. The number of European and American spouses is very substantial: 52 per 124, i.e. 42%, a high exogamy. However, in the Eskimo population in 1970 only 2% of genes were from these foreigners. This is due to the fact that the children born to mixed marriages were not assimilated, and as a result most of them had to emigrate.

The same exclusion phenomenon is to be found among the Kel Kummer Tuareg of Mali, who make up a very closed group of 370 individuals. This isolate was founded in the seventeenth century. The genealogical history, known since then, was reconstituted by Chaventré (1972) who published for this ethnic group a complete historical, sociodemographic and anthropobiological study (Chaventré, 1983). The genealogical reconstitution is based on 1167 individuals. On the whole there were 52 founders, original founders and immigrant founders. For a period of three hundred years, this number is very low. That could mean that the mating was almost exclusively between members of the same group. Effectively, for the last generation, the contribution of the fourteen principal founders belonging to the first and second generations is still 85%. Thus, a woman founder arriving during the second generation, alone contributed almost 12% of the gene pool. The maintenance of the initial contribution can be explained also by the progressive weeding out of imported genes. There were, in the course of time, some marriages with members of other ethnic groups. But the descendants of these marriages were not well assimilated. For example, the genes of two female founders who immigrated in the period between 1840 and 1870 represented at that time 1% of the gene pool. These genes disappeared as the descendants of the two women later left the group. Therefore, by the conscious or unconscious rejection mechanism of the host population, the foreign genes lingered only temporarily in this population. These factors led to the almost perfect maintenance of the Kel Kummer gene pool. This is an exceptional situation, in which the population is not only a social isolate, but also a genetic isolate.

The example of a French Protestant isolate in Normandy (present-day population 271) shows that cultural isolation does not necessarily lead to genetic isolation (Segalen & Jacquard, 1974; Jacquard, 1977). That isolate was established during the Reformation. It has always been a small island in a Catholic ocean. Until very recently, mixed marriages were prohibited. For this and other reasons, and also because of a great homonymy of family names, this group considered itself, and was considered by its Catholic neighbours, as very closed and inbred. As a matter of fact, the study of marriages contracted since 1800 demonstrates that the shrinking of the local matrimonial market necessitated the importation of spouses from other Protestant communities further away in France and also in Jersey. The probabilities of origin of genes show that in each generation the gene pool was rejuvenated. In the most recent generation, only 19% of genes can be traced to those present in the group of 1800. Hence the fear of a progressive genetic impoverishment was ill-founded.

Vu Tien Khang and Sevin (1977) described the genetic history of another region of France in the Pyrenees Orientales since the middle of the eighteenth century, of four relatively isolated villages of Pays de Sault. The number of inhabitants is about 50 for Le Clat, the smallest and most isolated village. For the largest village, Rodome, the corresponding number of inhabitants is about 160. According to the analysis of marriages, until the middle of the last century these populations practised strong village endogamy. Besides, the matrimonial migrations between these villages were assisted until this time, by their geographic proximity, of about half an hour to an hour on foot. These privileged relations have since deteriorated, at the start of the rural exodus, which left Rodome as the centre of attraction for the other villages. At the same time, the number of foreign spouses in the Pays de Sault increased, particularly in Rodome, the least isolated community and less in Le Clat, which is the most remote spot.

The evolution of the gene pool confirms these observations. In the latest generation, at the end of a 233 year history, about 89% of the genetic stock of Rodome was new, that of Le Clat only 35%. The two other villages hold an intermediate position, both in geography and in new genes (64% and 66%). These results show the extent which a group of populations, from a limited area and whose socioeconomic structures are similar, can have such dissimilar genetic histories. The sizes of these populations, their varying degrees of isolation, the different amounts of migration, the need to search for a mate or the ease with which a spouse can be found in a broader social context, the role of differential fecundity, all these are factors which influence genetic evolution in different ways.

The last example is that of another village in the Pyrenees, Arthez d'Asson, dealt with in the first part of this article. This village of about 500 inhabitants is the first in a mountain valley called Ouzom. Higher in this valley, there are two other villages, Arbeost and Ferrieres (respectively 190 and 200 inhabitants). The Ouzom valley is a well-defined geographical entity, and its three constituent populations are very close to each other. The geography and similar economic conditions seem therefore to have favoured frequent contact among all the inhabitants.

A preliminary study of the marital exchange over more than two centuries (Jacquard et al, 1975) on the contrary demonstrated the weakness of spouse-exchange between Arthez d'Asson and the two other villages. This phenomenon is connected with the historic conditions of settlement in the valley. Arthez d'Asson, which was an important hamlet of another commune (Asson) situated in the plain six kilometers away, became an independent community in 1744 as a result of its demographic expansion. The two other villages were established at the same time by founders who came from a nearby valley belonging to different cultural surroundings outside Bearn.

As the history of settlement would suggest, the POG show (Serre & Jakobi, 1985) that the contribution of spouses coming from Arbeost and Ferrieres to the gene pool of Arthez d'Asson at each generation is extremely small: it ranged from 0.5 to 2% and none at all during the last generation. The "new" genes come, on the one hand principally from the mother commune Asson and from another neighbouring village, and on the other hand and equally, from more distant regions (Bearn and its environs). Nevertheless, the new infusion is modest. In the latest generation, after a history of almost three centuries, more than 2/3 of the genes are "old" genes. This results in a pronounced genetic isolation in the eighteenth and nineteenth centuries, and one which is still increasing in the twentieth century. Several reasons can be advanced to explain this isolation. The first has to do with a question of method. The ascendancy of the spouses, originating from neighbouring communities, was also taken into account; and quite often the ancestors of these spouses were originally from Arthez d'Asson. In contrast to what happened with the Kel Kummer, the genes exported by inhabitants of Arthez d'Asson who emigrated and married into the neighbouring communities returned a few generations later with spouses brought to Arthez d'Asson. Another factor is to be found in the final differential fertility of the endogamous or exogamous couples. Eventually the drop in birthrate (Table 3) and the rural exodus (later in this region than in the Pays de Sault) limited the size of the population, and thus the number of eligible mates. But this limitation was not enough to make the inhabitants search for a spouse in more remote areas. In this case, such spouses would have had a modest chance of having ancestors from Arthez d'Asson.

The comparison of these few populations in terms of mate choice shows the extent to which particular circumstances and patterns of behaviour can have diverse consequences. The immigration of foreign spouses, substantial with the Eskimos, plays a small part in the composition of their gene pool. The immigrants, very few in the Jicaque Indians, have on the contrary made a very important contribution. The cultural isolation of the Protestant isolate in Normandy has not impeded genetic renewal at each generation. This isolation, in the case of the Kel Kummer Tuareg, has helped the diffusion of genes from a handful of founders. The interference in the traditional choice of a mate has led to a different kind of genetic renewal with respect for the four villages of the Pays de Sault. By restricting its choice to a geographically limited range, Arthez d'Asson preserved its genetic isolation.

CONCLUSION

In small isolated populations, economic and demographic equilibria are generally precarious and difficult to maintain. The rules of marriage can thus be thought of as a strategy to conserve such equilibria. They serve different aims: the preservation of landed property within the group, the regulation of the population size through exclusion of certain elements, the setting up of norms of alliance providing the best economic exchanges. Nevertheless, once certain geographic, demographic or socio-economic constraints - generally constraints - generally inevitable - are taken into account, such "adaptationalist" or "finalist" explanations do not always stand up to careful and detailed scrutiny. In fact, there is often a lack of consistency between what is supposed to be the norm for alliances and what is actually happening.

The disharmony can also appear when one analyses the evolution of the gene pool. Certain populations are under the illusion that in order to preserve their isolation, all they have to do is to limit the entry of foreign spouses; they are not aware that the gene pool may undergo a profound transformation as a result of only a few immigrants. Conversely, other populations, presumably exogamous, slow down within themselves the diffusion of foreign genes. These behaviour traits, belonging to populations as different as the Tuareg and the Eskimos, the Indians and some small western groups, show that the relationships between the history of individuals and the history of genes are not always obvious.

REFERENCES

Barrai, I., Cavalli-Sforza, L.L. & Moroni, A. (1962). Frequencies of pedigrees of consanguineous marriages and mating structure of the population. Annals of Human Genetics, **25**, 347-376.

Boyce, A.J., Küchemann, C.F., Harrison, G.A. (1967). Neighbourhood knowledge and the distribution of marriage distances. Annals of Human Genetics, **30**, 335-338.

Cavalli-Sforza, L.L. (1963). The distribution of migration distances, models and application to genetics. Entretiens de Monaco en Sciences humaines, Premiere Session, 24-29 mai, 1962. Human Displacements, pp. 139-158.

Cavalli-Sforza, L.L. & Hewlett, B. (1982). Exploration and mating range in African Pygmies. Annals of Human Genetics, **46**, 257-270.

Cazes, M.H. (1981). Les échanges matrimoniaux chez les Dogons de Tabi. Population, 1069-1084.

Chapman, A. & Jacquard, A. (1971). Un isolat d'Amérique Centrale: les Indiens Jicaques du Honduras. In: Génétique et Populations. Hommage à Jean Sutter. Population (Paris: INED-PUF), **60**, 163-185.

Chaventré, A. (1972). Un isolat du Sud-Sahara, les Kel Kummer. Population, 4-5, 769-783.

Chaventré, A. (1983). Evolution anthropo-biologique d'une population touarègue. Les Kel Kumer et leurs apparentés. Population (Paris: INED-PUF), **103**, 334.

Dahlberg, G. (1948). Mathematical Methods for Population Genetics. Basle: Karger. New York: Interscience.

Fix, A.G. (1974). Neighbourhood knowledge and marriage distance: the Semai case. Annals of Human Genetics, **37**, 327-332.

Gessain, R. (1970). Ammassalik ou la civilisation obligatoire, p. 252. Paris: Flammarion.

Hajnal, J. (1963). Concepts of random mating and the frequency of consanguineous marriages. Proceedings of the Royal Society (London), Series B, **159**, 125-177.

Jacquard, A. (1970). Structures Génétiques des Populations. Paris: Masson.

Jacquard, A. (1973). Distances généalogiques et distances génétiques. Cahiers d'anthropologie et d'écologie humaine, **1**, 11-123.

Jacquard, A. (1975). Inbreeding: one word, several meanings. Theoretical Population Biology, **7**, 338-363.

Jacquard, A. (1977). Concepts en Génétique des populations, p. 128. Paris: Masson.

Jacquard, A., Fernet, P. & Jakobi, L. (1975). Mariages et filiations dans la vallée pyrénéenne de l'Ouzom depuis 1744. Population, November 1975 (special issue), 187-196.

Malécot, G. (1948). Les Mathématiques de l'Hérédité. Paris: Masson.

Pison, G. & Lathrop, M. (1982). Méthode statistique d'étude de l'endogamie. Application à l'étude du choix du conjoint chez les Peul Bande. Population, **3**, 513-542.

Robert, J. (1971). Les Amassalimiut émigrés au Scoresbysund. Etude demographique et socio-économique de leur adaptation (cote orientale du Groenland, 1968). Bulletins et Mémoires de la Société d'Anthropologie de Paris, **8**, 5-135.

Roberts, D.F. (1968a). Differential fertility and the genetic constitution of an isolated population. Proc. 8th International Congress of Anthropological Sciences, **1**, 350-356.

Roberts, D.F. (1968b). Genetic effects of population size reduction. Nature, **220**, 1084-1088.

Ségalen, M. & Jacquard, A. (1974). Isolement sociologique et isolement génétique. Population, **28**, 551-570.

Serre, J.L. (1985). La consanguinite dans Deux Isolats Francais et Africains: Aspects Theoriques et Methodologiques de sa Mesure et de son Estimation. Doctoral thesis, Universite Pierre et Marie Curie, Paris.

Serre, J.L. & Jakobi, L. (1985). A genetic isolate in the French Pyrenees: Probabilities of origin of genes and inbreeding. Journal of Biosocial Science, **17**, 405-414.

Sutter, J. (1958). Evolution de la distance séparant le domicile des futurs époux. Population, **2**, 227-258.

Vu Tien Khang, J. & Sévin, A. (1977). Choix du conjoint et patrimoine génétique. Etude de quatre villages du Pays de Sault de 1740 à nos jours, p. 159. Paris: CNRS.

PART II

MATE CHOICE AND ASSORTATIVE MATING

ASSORTATIVE MATING
FOR PSYCHOMETRIC CHARACTERS

C. G. N. MASCIE-TAYLOR

Department of Biological Anthropology, University of Cambridge, England

INTRODUCTION and DEFINITIONS

Mate selection is a major biological event and its outcome is a substantial determinant of an individual's return on his/her whole reproductive investment (Lott, 1979). It is therefore hardly surprising that considerable interest has been devoted to the study of mate choice and in particular the effects of random and non-random mating. This chapter is devoted to the consideration of the effects of assortative mating for psychometric characters, particularly IQ and personality.

Assortative mating has been defined in various ways. Some geneticists use assortative mating in a "broad" sense to indicate any departure from random mating (panmixia) - consequently consanguineous mating would be part of assortative mating (Jacquard, 1970). Others reserve assortative mating to unions involving phenotypically similar individuals who are not relatives (Pennock-Roman, Vandenberg & Mascie-Taylor, 1988). The latter definition is found most often.

Assortative mating can be either positive (sometimes called homogamy) - where phenotypically like preferentially mate with like - or negative (heterogamy) where opposites preferentially mate (Garrison, Anderson & Reed, 1968). There is very little evidence for negative assortative mating in man with the possible exception of red hair colour in Europeans and albinism in the San Blas Indians of Central America (Stern, 1973). Consequently, human assortative mating implies, for the most part, positive mating and all future references to assortative mating will follow this general rule. Other terms also exist including endogamy, exogamy, mesalliance and hypergamy (see Eckland, 1968, and Epstein & Guttman, 1984, for definitions).

Some workers have also differentiated between assortative marriage and assortative mating. Garrison, Anderson and Reed (1968) used "marriage" rather than "mating" because they interpreted mating in conjunction with offspring, whereas many marriages are childless. Other workers have followed this usage (Mascie-Taylor & Gibson, 1979). Zonderman et al (1977) gave a different meaning to assortative marriage and mating by defining the latter as the phenotypic correlation between mates assuming no common environmental contribution. Assortative marriage is thus defined by them as including the common environmental contribution as well as the effects of convergence (the increasing similarity between spouses as a result of years of marriage). Since direct evidence for convergence has not been collected (see below, and also Epstein & Guttman, 1984), this paper uses the original definition put forward by Garrison et al (1968).

GENETIC EFFECTS OF ASSORTATIVE MATING

The genetic effects of assortative mating can be determined for both single gene and polygenic characters. There is little evidence of assortative mating for characters determined by single genes in man, although the theoretical aspects have been comprehensively discussed by Cavalli-Sforza and Bodmer (1971) and Jacquard (1970). The relationships between assortative mating and inbreeding have also been considered (Crow & Kimura, 1970; Crow & Felsenstein, 1968; Thiessen & Gregg, 1980). There are five potential effects of assortative mating for polygenic characters: (1) no change in gene frequencies will occur; (2) the frequency of genotypes giving rise to more extreme phenotypes will increase; (3) there will be a concomitant decline in heterozygosity; (4) there will be an increase in population variance; (5) parent-child and sib-sib correlations will increase. Crow and Felsenstein (1968) argued that the increase in homozygosity will tend to be small unless there is a very high level of assortative mating for the character and few gene loci involved. Even so the genetic structure of a population will be expected to be modified under this mating pattern, e.g. if the spouse correlation is 0.25 (and assuming certain equilibrium conditions), the variance in height would be expected to be one third greater than it would be under random mating (Crow & Kimura, 1970).

If assortative mating is accompanied by differential fertility then changes in gene frequencies can occur. Eckland (1968, 1972) suggested that assortative mating might be elevated in couples with high and low levels of intelligence; if this elevation also associates with differential fertility then there would be natural selection in favour of extreme levels of intelligence. Assuming that intelligence has some significant heritable component (Vernon, 1979; Bouchard & McGue, 1981), genetic variance would increase as a result of the combined action of assortative mating and differential fertility.

MATE CHOICE AND SELECTION

Evidence going back over one hundred years demonstrates that a husband and wife tend to resemble each other in a wide range of features including age, race, religion, ethnic origin, education, social class, smoking habits, values, physical characteristics including degree of pigmentation, aptitudes, personality and possibly psychopathology.

There are a large number of theories which have been put forward to try to explain mate choice and selection. The reasons for assortative mating have been reviewed elsewhere (see Epstein & Guttman, 1984; Pennock-Roman et al, 1986). Epstein and Guttman divided the theories into sociological, socio-biological and psychological groups. Pennock-Roman, Vandenberg and Mascie-Taylor (1986) provide a more historical overview and suggest that race, religion and education comprise the strongest boundaries defining the pool of eligible mates; within these boundaries persons will seek mates who are not too dissimilar in age (plus or minus 8 years). None of the theories (with the exception of Eckland's opportunity model) has attempted an explicit explanation for homogamy for aptitudes and intelligence.

EMPIRICAL EVIDENCE FOR ASSORTATIVE MATING

The earliest systematic study of assortative mating was done by Galton (1880) on the psychological and physical characteristics of famous English couples. Later studies examined a wide range of variables including social and geographical propinquity, educational similarity, physical features such as weight and height, IQ, personality, ethnic background, smoking and drinking habits, and even alliteration - partners whose first name begin with the same letter (for reviews see Vandenberg, 1972; Spuhler, 1968; Thelen, 1984; Roberts, 1977; Jensen, 1978; Johnson 1980; Kopelman & Lang, 1985). Most studies have found a tendency for "like to marry like". Indeed, Murstein (1980) concluded that "assortativeness for marriage continues to be strongly evident with no variable having been shown to be completely independent." The majority of the studies of both general and specific abilities are based on American or European white samples, although Johnson et al (1976a) also found positive assortative mating for cognitive abilities in a Korean sample.

Figure 1 provides a summary of all spousal similarities where an overall test of intelligence was given. The first results were reported nearly 60 years

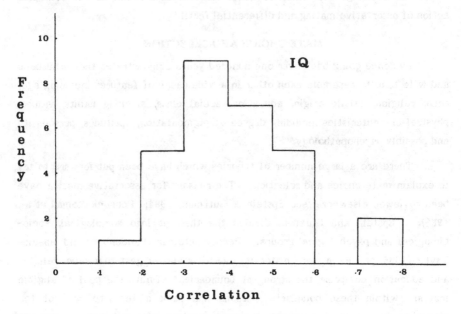

Figure 1:   Example of assortative mating for IQ.

ago.   Jensen reviewed the studies up to 1978 and found a median spousal correlation of 0.44 with a range from 0.12 to 0.76.   Bouchard and McGue (1981) found a weighted mean correlation of 0.33 based on 3817 couples, although their sample is more selective than Jensen's.   Pennock-Roman, Vandenberg and Mascie-Taylor (1988) provide information on more recent studies as well as those omitted by Jensen.

Specific cognitive abilities have been examined less frequently, and there is considerable variability in the correlations.   In general, verbal tests tend to show the highest spousal association and spatial tests the lowest.

Reviews of personality traits are to be found in Spuhler (1967), Vandenberg (1972), Murstein (1976), Kephart (1977) and Pennock-Roman et al (1986). In general there is some evidence of at most low to moderate homogamy (Figure 2).   However, as noted by Pennock-Roman et al, research in this area has been characterised by small sample sizes and poor tests (Anastasi, 1976). The most cited results are those of Cattell and Nesselroade (1967) who grouped marriages as "stable" or "unstable".   Significant disassortative mating was found for several traits in the unstable group including outgoingness (-0.50), enthusiasm (-0.40), self-sufficiency (-0.32) and suspicious, self-opinionated (-0.33).

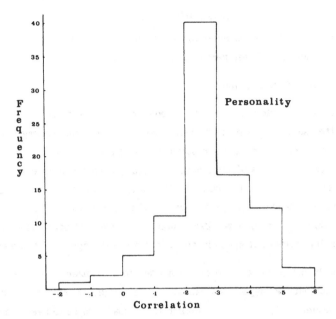

Figure 2:   Example of assortative mating for personality traits.

Table 1:   Effects of Convergence for IQ and Personality

Correlation between years of marriage and difference in IQ scores and personality between spouses.

	Cambridge Sample (N=193)		Oxford Sample (N=150)	
WAIS IQ Test	r	p	r	p
Comprehension	-0.189	NS	+0.116	NS
Similarities	-0.015	NS	+0.046	NS
Vocabulary	+0.014	NS	-0.005	NS
Digit Span	-0.171	NS	+0.036	NS
Block Design	+0.003	NS	-0.123	NS
Object Assembly	-0.042	NS	+0.036	NS
Digit Symbol	+0.020	NS	+0.029	NS
Verbal IQ	-0.132	NS	+0.105	NS
Performance IQ	-0.055	NS	-0.008	NS
Total IQ	-0.106	NS	+0.035	NS
EPQ				
Extraversion	+0.015	NS	+0.153	NS
Neuroticism	+0.117	NS	-0.028	NS
Inconsistency	+0.096	NS	+0.137	NS

CONFOUNDING EFFECTS

A number of factors need to be taken into account in interpreting the empirical findings for IQ and personality.

(a) *Effects of Convergence*

The first problem arises due to convergence - the degree of similarity present after some years of marriage. As Price and Vandenberg (1980) note, most studies measure realised assortment which is the degree of similarity present after some years of marriage. They argue that the observed similarity between spouses could be due to (1) initial similarity during courtship, (2) convergence during marriage as a result of shared environments or reciprocal influence, (3) attrition of dissimilar couples as a result of separation or divorce, (4) confounding effects of age related variables in heterogeneous samples.

Ideally, longitudinal studies are needed to test directly for the effects of convergence, but the hypothesis can be tested indirectly by examining whether couples married longer are more similar than those who have been married only briefly.   Most studies that have looked for the effects of convergence have found little or no evidence to support it (Harrison et al, 1976; Mascie-Taylor & Gibson, 1979; Zonderman et al, 1977).   However, as Price and Vandenberg point out, these studies have removed the linear effects of age or length of marriage but this does not test for the occurrence of phenotypic convergence.   They recommend using hierarchical multiple regression in which the dependent variable is, for example, wife's IQ score.   The years of marriage are entered as the first step, husband's IQ score next, and finally the interaction term.   The interaction term is then independent of the other effects.   Even so, Price and Vandenberg (1980) failed to demonstrate any convergence except for "highly plastic variables such as alcohol consumption and amount of social activity." Re-analysis of the Otmoor (Harrison et al, 1974) and Cambridge (Mascie-Taylor, 1977; Mascie-Taylor & Gibson, 1979) data using the interaction method (Mascie-Taylor, 1988) provided no consistent evidence for similarity accruing during marriage for IQ or personality traits (Tables 1 & 2).   Buss (1984) reported consonant results for his self-reported personality measures.

A different approach was used by Watkins and Meredith (1981) who examined newly-wed couples tested on 21 tests of specific cognitive abilities. Fifteen of these tests had been administered previously in the Hawaian Family Study (Johnson et al, 1976) and the Colorado Study (Zonderman et al, 1977) where most couples had been married for many years (although exactly how long is unclear in the Colorado Study).   All the Hawaiian couples had teenage

Table 2: Interaction of Spouse Similarity with Years of Marriage

	Hierarchical Step	Cambridge			Oxford		
		r	R^2	p	r	R^2	p
WAIS IQ Test							
Comprehension	Years of marriage	0.21	0.04	NS	0.13	0.02	NS
	Husband's score	0.28	0.08	<0.001	0.32	0.10	<0.001
	Interaction	0.33	0.10	NS	0.34	0.12	NS
Similarities	Years of marriage	0.01	0.00	NS	0.04	0.00	NS
	Husband's score	0.40	0.16	<0.001	0.44	0.20	<0.001
	Interaction	0.40	0.16	NS	0.45	0.20	NS
Vocabulary	Years of marriage	0.09	0.01	NS	0.21	0.04	NS
	Husband's score	0.37	0.14	<0.001	0.25	0.06	NS
	Interaction	0.37	0.14	NS	0.26	0.07	NS
Digit Span	Years of marriage	0.00	0.00	NS	0.08	0.01	NS
	Husband's score	0.22	0.05	<0.001	0.41	0.17	<0.001
	Interaction	0.24	0.06	NS	0.42	0.18	NS
Block Design	Years of marriage	0.09	0.01	NS	0.55	0.30	<0.001
	Husband's score	0.21	0.04	<0.01	0.58	0.34	<0.01
	Interaction	0.21	0.05	NS	0.58	0.34	NS
Object Assembly	Years of marriage	0.13	0.02	NS	0.33	0.11	<0.001
	Husband's score	0.28	0.08	<0.01	0.34	0.11	NS
	Interaction	0.29	0.08	NS	0.34	0.12	NS
Digit Symbol	Years of marriage	0.09	0.01	NS	0.24	0.06	NS
	Husband's score	0.29	0.08	<0.001	0.26	0.07	NS
	Interaction	0.30	0.09	NS	0.26	0.07	NS
Verbal IQ	Years of marriage	0.03	0.00	NS	0.14	0.02	NS
	Husband's score	0.35	0.13	<0.001	0.47	0.22	<0.001
	Interaction	0.36	0.13	NS	0.47	0.23	NS
Performance IQ	Years of marriage	0.04	0.00	NS	0.12	0.02	NS
	Husband's score	0.28	0.08	<0.001	0.21	0.04	<0.05
	Interaction	0.29	0.08	NS	0.23	0.05	NS
Total IQ	Years of marriage	0.07	0.00	NS	0.02	0.00	NS
	Husband's score	0.38	0.15	<0.001	0.37	0.14	<0.001
	Interaction	0.38	0.15	NS	0.38	0.14	NS
EPQ							
Extraversion	Years of marriage	0.05	0.00	NS	0.09	0.00	NS
	Husband's score	0.24	0.06	<0.01	0.16	0.03	NS
	Interaction	0.25	0.06	NS	0.21	0.04	NS
Neuroticism	Years of marriage	0.04	0.00	NS	0.09	0.01	NS
	Husband's score	0.09	0.01	NS	0.10	0.01	NS
	Interaction	0.10	0.01	NS	0.21	0.02	NS
Inconsistency	Years of marriage	0.15	0.02	<0.05	0.19	0.04	NS
	Husband's score	0.25	0.06	<0.01	0.31	0.09	<0.01
	Interaction	0.25	0.06	NS	0.34	0.11	<0.05

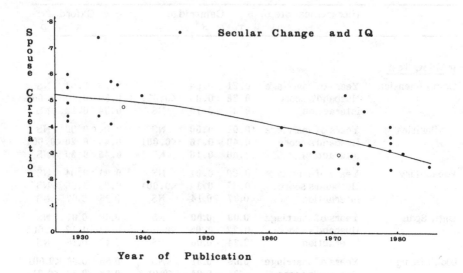

Figure 3:   Temporal decline in levels of assortative mating for IQ.

children.   Comparisons between Watkins and Meredith's results and the other two studies revealed no consistent differences which suggests that the similarity between couples is due to initial assortment.

It would seem that most of the available indirect evidence supports the notion of initial similarity between couples for IQ and personality factors. Convergence only seems to be relevent for "plastic variables" such as smoking and drinking behaviour, and Price et al (1982) suggest spouses assorting for similarity in smoking reinforce this behaviour afterwards thus providing a strongly conservative influence.

(b)   *Effect of secular trend*

Johnson et al (1980) suggested that there might be a secular trend in the degree of assortative mating for IQ, with higher levels before 1946 (correlation coefficient 0.48) than in recent times (0.30; see Figure 3).   They put forward several reasons for the apparent decline in homogamy: less representative samples were tested earlier with unstandardised tests compared with more recent studies; recent tests tend to measure spatial ability which shows lower spousal association than more verbal tests.   As noted earlier, verbal tests show higher levels of homogamy.   They do not rule out a real decline in homogamy for IQ.

Watkins and Meredith (1981) felt that the evidence for the secular trend in homogamy for intelligence was not so clear cut since the distribution of dates of marriage for most couples was not reported.   Harrison et al (1976) found few systematic trends although the correlation for couples married between 1940 and 1949 was significantly lower than for those married after 1950 as well as for those married prior to 1940.   These results are in general accord with the views of Watkins and Meredith.   They propose that the lowest level of spousal IQ similarity would be at the time of the second World War when educational homogamy was also low.   Using a six point scale for educational class, Rockwell (1976) showed that from 1931 to 1970 the percentage of white people marrying within their own class had hardly varied but couples marrying between 1941 and 1945 had the lowest percentage of educational class homogamy.

Since IQ and education are highly correlated (Vernon, 1979), the contribution of educational homogamy must clearly be considered.

(c)   *Effect of education and social class (socioeconomic status)*

(1)   Educational similarity between spouses:

The results from the 1960 and 1970 U.S. Censuses (Michielutte, 1972; Rockwell, 1976) have consistently shown that education has been, and continues to be, an important factor in marital choice.   In Britain, Mascie-Taylor (1986b) found age at leaving school was similar in over 7600 couples who formed part of a randomly selected cohort.   Of those husbands who continued into higher (further) education, 82% had wives who had stayed beyond the minimum school leaving age with some 64% also continuing to higher education.   These results are in accord with findings which show that in spouses who differ in extent of education it is commoner for the husband to be more educated.   The degree and pattern of spouse resemblance for education in France is very similar to that in the United States of America and Great Britain (Girard, 1964).

The secular change in educational homogamy has already been mentioned. In a 5% national sample from the 1970 Census (n = 443,520 marriages), the percentage of white couples in which both spouses had the same amount of education (measured on a 5-point scale) dropped from 70.7% for the cohorts married in 1910 or earlier to 42.3% in 1941-1945 cohorts.   From 1946 to 1970 there was a levelling off at around 46.0%.   Rockwell attributes this decrease to increased educational heterogeneity rather than diminution in the desire to marry homogamously.   Pennock-Roman, Vandenberg and Mascie-Taylor (1986) explain the effects of educational heterogeneity as follows: "If everyone was illiterate, heterogeneity would be zero and everyone would marry homogamously for education (by chance or otherwise). With the increase in the number of persons attaining higher education throughout this century, the distribution of amount of education has become more heterogeneous, which increases the chances for heterogamy."

Warren (1966) also reported a trend in educational homogamy in a U.S. national sample of 33,000 households interviewed in 1962. He computed husband/wife correlations for education by age groups, using a nine point scale. Non-white couples showed a wider range of correlations (0.39 to 0.70) than white couples (0.55 to 0.63) but the lowest level of homogamy in whites appeared to be in the group married between 1941 and 1945 which would agree with Rockwell's report.

Recently, Ahern et al (1983) suggested that there might be a generational difference in assortative mating for educational attainment in three major ethnic groups (persons of Chinese, European and Japanese ancestry) who took part in the Hawaiian Family Study. The sample comprised parents and grandparents. The mean age of the parents was known (45 years for males and 43 for females) and the grandparents were probably born in the period 1900 to 1909. All ethnic groups showed a decline in assortative mating for years of education from grandparental to parental generation. The Europeans showed a fall from 0.58 to 0.44, Japanese 0.72 to 0.39, and Chinese 0.59 to 0.42. Only the Chinese decline was insignificant, even though the magnitude of the decline was greater than that found in the European ancestry group. The lack of significance can partly be ascribed to the small size (n = 84) of the Chinese ancestry sample.

Heath et al (1985) examined Norwegian twin data and estimated the marital correlation for recalled educational achievement as 0.67 in the twins and their spouses, and 0.80 in the parents of twins and their spouses – a result in keeping with Ahern et al. However, they argue that the difference is an artefact due to bias in recall of parental educational level dependent on the twins' own educational level. The estimates of Heath et al are substantially higher than those reported by Ahern et al, possibly because they use "polychoric" (Olsson, 1979) correlations rather than treating educational level as a continuous variable.

(2) Association between IQ, social class and education:

Eckland (1970) suggested that the major reason for assortative mating for intelligence and other abilities was because of the strong tendency to marry someone of similar education. As we have already seen, educational homogamy is particularly marked. Johnson et al (1976) using age-adjusted tests found "near zero" spouse correlations after educational level had been partialled out - suggesting that there is very little spouse similarity for that part of cognitive ability which is independent of attainment. They used partial correlation analysis to statistically adjust cognitive ability for educational level, although this procedure would "overcorrect the spouse correlation". Even so, the residual correlations for Perceptual Speed and Verbal factor scores were both 0.12 and statistically significant.

In Western societies there is a positive association between higher social class or socioeconomic status and higher IQ, consequently some of the observed IQ similarity between spouses could be accounted for by social homogamy. In addition, social class and education are also highly correlated.

Early studies by Centers (1949) and Hollingshead (1950) showed a significant tendency for spouses to marry within their own class or one step below or above. Warren (1966) in the national sample found endogamy for class using fathers' occupations for both sexes to be 55% based on a three point classification (white and blue collar and farm). Chase (1975) re-analysed the same data by age groups and used the husband's first job with the wife's original (father's) class and found little change in endogamy between 1921 and 1960.

Because of the high correlation with education, Warren (1966) re-analysed the national sample and found that the spouse correlation for socioeconomic status (of origin) fell by one half when the effects of educational level were statistically controlled. Blau and Duncan (1967) interpreted these results as follows, "It is plausible to think of assortment as occurring directly and primarily on the basis of schooling and personal characteristics of the mates and only secondarily and indirectly on the basis of the occupational status of the parental generation."

(3)   Extent of assortative mating after partialling out social class, educational and other effects:

Watkins and Meredith (1981) examined educational level, income (parental and own), social class and occupational prestige in their newly-wed sample in relation to cognitive scores. They found the husband/wife correlation for the Verbal/Reasoning factor fell from 0.4 to 0.3 after partialling out education. After partialling out eight other socioeconomic variables, pertaining to one or other spouse, the correlation fell slightly further, but even so the residual correlation was still significant. Spousal correlations for Spatial and Memory factors were unaffected by partialling out education and the socioeconomic variables but only the Spatial factor was significant.   Warren and Meredith also performed a multivariate test for spousal similarity in socioeconomic status with all variables taken together.   The canonical correlation was not much higher than for own education alone, suggesting "that parental social class variables play a small role in marital choice, independently of education."

Mascie-Taylor and Vandenberg (1988) using data from a Cambridge (UK) family study considered the extent to which the similarity between 193 spouses for IQ and personality could be "explained" by social, educational, geographical and familial propinquity.   They attempted "to differentiate the broader effects of propinquity from specific mate selection based on personal preference *per se*." Consequently, they attempted to differentiate between active and passive elements of mate selection and to determine their relative contribution to assortative mating.   Both partial correlation and stepwise multiple regression analyses were used.   For IQ subtests and IQ components of the Wechsler Adult Intelligence Scale, the passive elements defined as those resulting from educational, social and geographical propinquity accounted for approximately two-thirds of the total association.   However, the remaining third ascribed to personal choice was still significant.   For personality (using Eysenck's Personality Inventory) the overall assortative mating levels were low, but even so the personal choice component was still significant.   Personality played no role in mate similarity for IQ or vice versa.

Mascie-Taylor et al (1986) employed a slightly different statistical approach when re-analysing assortative mating for IQ (Wechsler Scale) in 150 husband/wife pairs from the Otmoor Study, Oxfordshire (Harrison et al, 1974; Harrison et al, 1976).   Using stepwise multiple regression the effects of years of education, social class, type of school, geographical propinquity (defined simply as local or non-local), birth order and family size were removed independently (cf. Mascie-Taylor & Vandenberg, 1988) for husbands and wives.   The residual spousal correlations after removal of these "independent variables" were computed.   Of the IQ components only Verbal IQ showed a significant association;   for the subtests, vocabulary was significant at the 5% level and similarities and block design at the 10% level (Table 3).   These results are in broad agreement with those of Mascie-Taylor and Vandenberg (1988) where the same IQ test was used.

Heath and Eaves (1985) have also put forward ways of resolving the effects of phenotype and social background on mate selection.   Their approach is based on an extension of the classical twin design (Eaves, 1970; Martin et al, 1978).   They suggest that collection of data on monozygotic and dizygotic twin pairs and their spouses together with the estimation of all possible correlations

Table 3:   Effects of homogamy on IQ correlations (from Mascie-Taylor et al, 1988)

Wechsler IQ Test	Zero order correlation	p	After removal of independent variables[1]	p
Comprehension	+0.287	<0.001	+0.06	NS
Similarities	+0.442	<0.001	+0.13	<0.10
Digit span	+0.097	NS	+0.03	NS
Vocabulary	+0.406	<0.001	+0.15	<0.05
Block design	+0.321	<0.001	+0.12	<0.10
Object assembly	+0.162	<0.05	+0.06	NS
Digit symbol	+0.130	NS	+0.01	NS
Verbal IQ	+0.462	<0.001	+0.16	<0.05
Performance IQ	+0.155	NS	+0.06	NS
Total IQ	+0.372	<0.001	+0.09	NS

[1] Independent variables were years of education, type of school attended, family size, birth order, social class, and geographical propinquity.

between twins and spouses will resolve this issue. They are particularly concerned with determining whether positive assortative mating is the result of "phenotypic" similarity which is implicit in most behavioral genetic analyses (Rice et al, 1980) or whether it arises from similarity in social background. They had previously argued (Heath & Eaves, 1984) that the effects of social background on mate selection will have different implications for genetic analysis from those of phenotypic assortative mating. As they point out, their mixed homogamy model using path analysis (Heath & Eaves, 1984, 1985) does not exhaust all possible models (Cloninger, 1980) nor has it been tested. The majority of studies which have partialled out education and/or social class, etc., find a large decline in the observed spousal correlation for IQ and personality. However, in at least four studies significant residual spousal correlations remain particularly for verbal ability. Whether this implies personal preference or active choice remains to be seen - there are always "background variables" not considered which could reduce the residual spousal correlation (Mascie-Taylor, 1986a).

ASSORTATIVE MATING AND FERTILITY

(a)   *Direct evidence*

Assortative mating, if associated with differential fertility, can bring about evolutionary changes in the genetic structure of populations. There are, however, very few studies which have examined this relationship directly and those that do consider anthropometric rather than psychometric characters.

Spuhler (1967) has been incorrectly reported (Thiessen & Gregg, 1980) as having shown that the degree of assortment for IQ positively associates with fertility. This confusion might have arisen because Spuhler (1968) reported on this relationship for anthropometric characters (see below). In the 1967 paper, Spuhler reported assortative mating for two intelligence tests, the Progressive Matrices and the verbal meaning part of the Progressive Mental Abilities test. The correlations were 0.399 and 0.305 respectively. (Both correlations refer to "total right". If proportion right is used the Matrices correlation increases to 0.732.)

Spuhler also related test scores to fertilty. Since many couples in the Ann Arbor sample had not completed their reproductive period, Spuhler computed a fertility index. The square-root transformation of months of exposure to pregnancy plotted against the number of live-born children produced a strong approximately linear relationship between exposure and fertility. Using this, a score was obtained giving the difference between observed and expected fertility for each couple. This fertility score was normally distributed with a mean of 0.0003 and standard deviation of 1.06. Positive correlations were obtained between mothers' fertility and their intelligence (Matrices 0.148, Mental Abilities 0.128 or 0.234), but they were not significant for fathers (0.032, -0.010 or 0.038 respectively).

In other papers, Clarke and Spuhler (1959), Spuhler (1962) and Spuhler (1968) reported on anthropometric mating with respect to differential fertility also using an Ann Arbor sample. Of 43 measurements, 14 showed no significant assortative mating. For the remaining 29 measurements Spuhler computed an index of similarity so as to relate similarity to fertility. The similarity index, the ratio of the husband's measurement to the sum of the husband's and wife's measurement, i.e. (H/H+W), was approximately normally distributed for most of the 29 measurements. The correlation between similarity and fertility (using the same fertility index as before) showed that only one correlation was significant (minimum wrist circumference = +0.175). Spuhler, however, only examined the linear association component whereas his similarity index would be expected to show a polynomial relationship with fertility (assuming some underlying relationship between spouse likeness and number of live-born offspring).

There are several measures of similarity which could be used. For instance, the absolute difference between spouses' scores (or measurements) would provide a crude estimate (assuming no mean sex differences) or the normalised difference which would correct for any disparity between means.

Table 4:   Relationship between fertility and similarity (from Mascie-Taylor, 1987)

| | Cambridge | | | | Oxford | | | |
| | Absolute | | Normalised | | Absolute | | Normalised | |
	r	p	r	p	r	p	r	p
Wechsler IQ Test								
Comprehension	-0.195	NS	-0.057	NS	+0.025	NS	+0.019	NS
Similarities	-0.109	NS	-0.094	NS	+0.108	NS	+0.114	NS
Vocabulary	+0.076	NS	+0.088	NS	+0.142	NS	+0.164	<0.05
Digit span	-0.001	NS	0.000	NS	-0.007	NS	-0.014	NS
Block design	-0.008	NS	+0.007	NS	-0.039	NS	-0.087	NS
Object assembly	-0.130	NS	-0.139	NS	-0.113	NS	-0.056	NS
Digit symbol	-0.096	NS	-0.088	NS	-0.059	NS	-0.045	NS
Verbal IQ	+0.006	NS	+0.036	NS	+0.087	NS	+0.079	NS
Performance IQ	-0.099	NS	-0.090	NS	-0.130	NS	-0.121	NS
Total IQ	-0.072	NS	-0.050	NS	-0.087	NS	-0.089	NS
EPI								
Extraversion	+0.006	NS	-0.006	NS	-0.079	NS	-0.005	NS
Neuroticism	+0.043	NS	+0.024	NS	-0.003	NS	+0.003	NS
Inconsistency	+0.107	NS	+0.102	NS	-0.081	NS	-0.091	NS

Mascie-Taylor (1988) have recently examined both Cambridge (Mascie-Taylor & Gibson, 1979; Mascie-Taylor & Boldsen, 1984) and Oxford (Harrison et al, 1974, 1976) IQ and personality data using both these measures and Spuhler's fertility index. In both populations there was a highly significant linear association between the square-root of the exposure to pregnancy plotted against the number of live-born children. The correlation between exposure and fertility was 0.40 and 0.38 in Cambridge and Oxford populations respectively. Polynomial curves of exposure against fertility provided no better fits than did the square-root of exposure.

The correlations of the fertility score (the difference between observed and expected fertility) with similarity measures are presented in Table 4. It can be seen that there is little or no association between increasing spousal similarity and increasing fertility for either IQ or personality traits in the Oxbridge samples. Indeed, the only significant association is with the vocabulary subtest in the Oxford sample and the positive correlation indicates that increasing dissimilarity for vocabulary associates with fertility score! Furthermore, one significant result at the 5% level can easily be accounted for by chance alone.

These studies do not provide any substance for the hypothesis of an association between increasing homogamy and fertility.   However, in both samples fertility is incomplete and the fertility score provides only an approximate measure of the true fertility.   Mascie-Taylor and Boldsen (1988) examined the relationship between spousal height similarity and completed family size as well as abnormal pregnancies in their British cohort study.   They found increasing abnormalities associated with increasing dissimilarity in height; completed family size was higher in couples who were more similar even after correcting for regional, social and parental age.

(b)   *Indirect evidence*

There is also indirect evidence from educational studies of a relationship between assortative mating and differential fertility.   Kiser (1968) studied the effect on fertility of assortative mating for education (using U.S. census data) by comparing observed fertility levels with those expected for random mating.   He found differences between whites and non-whites.   Assortative mating led to modest increases of fertility at most educational levels for whites but modest decreases for non-whites at most levels.   For both whites and non-whites assortative mating increased the fertility of the lowest educational group the most.

Garrison et al (1968) showed that positively assortative marriages for educational attainment resulted in the production of more children than negatively assortative marriages (Figure 1), mostly because a larger percentage of negatively assortative marriages are childless.

Mascie-Taylor (1986b) using a British national sample also examined the relationship between educational homogamy and fertility.   There were no childless couples since the sample centered on children born in a specific week of March 1958.   The average age for mothers at that time was 28 years.   The children and their families were re-studied periodically, in 1965, 1969, 1974 and in 1983.   The fertility data were collected in 1974 which would be close to completed fertility for most families since the average age of mothers would by then have been 44 years of age.   Education was measured as age at which the individual left full-time education.   The differences in years between husband's and wife's education were computed and related to the number of live born children.   Since there were few marriages with partners differing by more than three years, the educational homogamy was reduced to four categories of no difference (0) and 1 to 3 years.   Because negative and positive (-1 or +1, husbands with one year less or one year more education than their wives, respectively) were essentially symmetric, the sign was ignored.   Table 5

Table 5: Analysis of variance for fertility

Item	df	Mean square	F	p
Age of husband	1	1.864	0.662	NS
Age of wife	1	36.107	12.825	<0.001
Social class	4	196.852	69.921	<0.001
Educational homogamy	3	24.685	8.768	<0.001
Interaction class X homogamy	12	2.678	0.951	NS
Residual	7048	2.815		

Multiple Classification Analysis

		Unadjusted	Adjusted for spousal age
Educational homogamy/heterogamy	0	+0.13	+0.07
	1	-0.11	-0.04
	2	-0.31	-0.23
	3	-0.44	-0.20
Social class	I	-0.49	-0.42
	II	-0.36	-0.33
	III	-0.03	-0.04
	IV	+0.39	+0.36
	V	+1.17	+1.12

demonstrates that as non-assortative mating for education increased the average number of children per family declined.

Although these results are in agreement with the findings of Kiser and Garrison et al, there are potentially confounding variables which could account for the observed association. For instance, social class and family size are known to be associated, as are fertility and parental age. Since the national sample deals with a cross-section of all parents the ages of mothers and fathers vary. Older parents would be expected to have more children. If educational homogamy was higher in the lower social classes or the average parental ages differed between classes then a spurious association between educational homogamy and fertility would occur.

An analysis of variance of fertility removed the effects of parental ages, social class, educational homogamy, as well as the interaction between class and homogamy. The results (Table 5) show that the age on the mother (but not the father) associates very strongly with fertility. The multiple classification analysis indicated that fertility increased from social class I to V but within

Table 6:   Family size by social class and educational homogamy

Educational Homogamy/heterogamy	I	II	III	IV	V	Total
0	2.97	2.97	3.39	3.84	4.55	3.46
1	2.95	3.01	3.18	3.50	4.44	3.23
2	2.68	2.88	3.05	3.46	3.94	3.03
3	2.67	2.99	2.92	3.29	5.00	2.89
Total	2.84	2.97	3.30	3.73	4.50	3.33

each social class there was a trend for decreasing fertility as educational homogamy decreased (Table 6).   A stepwise multiple regression analysis which took into account linear and higher order effects and their interactions confirmed the finding of a strong relationship between degree of assortative mating for years of education and fertility.   Since there is a high correlation between educational level and IQ these results lend indirect support to the notion of a relationship between assortative mating and differential fertility.

The various studies cited provide clear evidence for positive assortative mating for IQ with lower correlation coefficients for personality traits. Indirect evidence does not support the idea that spouses become more similar with longevity of the marriage and one can assume that the observed associations occur prior to marriage and not as a result of years of marriage. However, only longitudinal studies will confirm this hypothesis.

The secular trend for assortative mating for IQ - with higher correlations before the last war and lower values afterwards - can most easily be explained by declining educational homogamy in more recent times, but the paucity of data, different IQ tests and variable age cohorts renders examination of this trend for IQ particularly difficult.

By far the most interesting feature from an evolutionary standpoint is whether there is an association between assortative mating and differential fertility for IQ - assuming IQ to have a significant heritable component. Very few studies have considered this topic and the results have tended to show no association.   However, the combination of small sample sizes and incomplete fertility makes firm conclusions difficult to reach whilst the indirect evidence using educational level instead of IQ lends strong support for a relationship between assortative mating and differential fertility.   Longitudinal studies of couples should provide answers to many of these unresolved problems.

ACKNOWLEDGMENTS

This research was supported by a grant from the Marie Stopes Research Fund of the Eugenics Society.

REFERENCES

Ahern, F.M., Johnson, R.C. & Cole, R.E. (1983). Generational differences in spouse similarity in educational attainment. Behavior Genetics, **13**, 95-98.

Anastasi, A. (1976). Psychological Testing (4th edn.). New York: MacMillan.

Blau, P.M. & Duncan, O.D. (1976). The American Occupational Structure. New York: Wiley.

Bouchard, T.J. & McGue, M.G. (1981). Familial studies of intelligence: a review. Science, **212**, 1055-1059.

Buss, D.M. (1984). Marital assortment for personality disposition: assessment with three different data sources. Behavior Genetics, **14**, 111-123.

Cattell, R.B. & Nesselroade, J.R. (1967). Likeness and completeness theories examined by 16 P.F. measures on stably and unstably married couples. Journal of Personality and Social Psychology, **7**, 351-361.

Cavalli-Sforza, L.L. & Bodmer, W.F. (1971). The Genetics of Human Populations. San Francisco: Freeman.

Centers, R. (1949). Occupational endogamy in marital selection. American Journal of Sociology, **54**, 530-535.

Chase, I.D. (1975). A comparison of men's and women's intergenerational mobility in the U.S. American Sociological Review, **40**, 483-505.

Clarke, A.C. & Spuhler, J.N. (1959). Differential fertility in relation to body dimensions. Human Biology, **31**, 121-137.

Cloninger, C.R. (1980). Interpretation of intrinsic and extrinsic structural relations by path analysis: theory and applications to assortative mating. Genetical Research, **36**, 133-145.

Crow, J.F. & Felsenstein, J. (1968). The effect of assortative mating on the genetic compposition of a population. Eugenics Quarterly, **15**, 85-97.

Crow, J.F. & Kimura, M. (1970). An Introduction to Population Genetics Theory. New York: Harper & Row.

Eaves, L.J. (1970). Aspects of Human Psychogenetics. Unpublished Ph.D. thesis, University of Birmingham, England.

Eckland, B.K. (1968). Theories of mate selection. Eugenics Quarterly, **15**, 71-84.

Eckland, B.K. (1970). New mating boundaries in education. Social Biology, **17**, 269-277.

Eckland, B.K. (1972). Evolutionary consequences of differential fertility and assortative mating in man. In: Th. Dobzhansky, M. K. Hecht & W. C. Steere (eds.), Evolutionary Biology, vol.5. New York: Appleton.

Epstein, E. & Guttman, R. (1984). Mate selection in man: evidence, theory, and outcome. Social Biology **31**, 243-278.

Galton, F. (1880). Hereditary Genius. New York: Appleton.

Garrison, R.J., Anderson, V.E. & Reed, S.C. (1968). Assortative marriage. Eugenics Quarterly, **15**, 113-127.

Girard, A. (1964). Le Choix du Conjoint. Paris: Presses Universitaires de France.

Harrison, G.A., Kuchemann, C.T., Hiorns, R.W. & Carrivick, P.J. (1974). Social mobility, assortative marriage and their interrelationships with marital distance and age in Oxford City. Annals of Human Biology, **1**, 211-223.

Harrison, G.A., Gibson, J.B. & Hiorns, R.W. (1976). Assortative marriage for psychometric, personality, and anthropometric variation in a group of Oxfordshire villages. Journal of Biosocial Science, **8**, 145-153.

Heath, A.C. & Eaves, L.J. (1984). Elements of a general linear model of assortative mating. Behavior Genetics, **14**, 108-114.

Heath, A.C. & Eaves, L.J. (1985). Resolving the effects of phenotype and social background on mate selection. Behavior Genetics, **15**, 15-30.

Heath, A.C., Berg, K., Eaves, L.J., Solaas, M.H., Sundet, J., Nance, W.E., Corey, L.A. & Magnus, P. (1985). No decline in assortative mating for educational level. Behavior Genetics, **15**, 349-369.

Hollingshead, A.B. (1950). Cultural factors in the selection of marriage mates. American Sociological Review, **15**, 619-627.

Jacquard, A. (1970). The Genetic Structure of Populations. Heidelberg: Springer-Verlag.

Jensen, A.R. (1978). Genetic and behavioral effects of non-random mating, In: C. E. Noble, R. T. Osborne & E. N. Weyl (eds.), Human Variation: The Biopsychology of Age, Race and Sex. New York: Academic Press.

Johnson, R.C., Deries, J.C., Wilson, J.R., McClearn, G.E., Vandenberg, S.G., Ashton, G.C., Mi, M.P. & Rashad, M.N. (1976). Assortative marriage for specific cognitive ability in two ethnic groups. Human Biology, **48**, 343-352.

Johnson, R.C., Ahern, F.M. & Cole, R.E. (1980). Secular change in degree of assortative mating for ability. Behavior Genetics, **10**, 1-8.

Kephart, W.M. (1977). Family, Society and the Individual. Boston: Houghton-Mifflin.

Kiser, C.B. (1968). Assortative mating by educational attainment in relation to fertility. Eugenics Quarterly, **15**, 98-112.

Kopleman, R.E. & Lang, D. (1985). Alliteration in mate selection: does Barbara marry Barry? Psychological Reports, **56**, 791-796.

Lott, D.F. (1979). A possible role for generally adaptive features in mate selection and sexual stimulation. Psychological Reports, **45**, 539-546.

Martin, N.G., Eaves, L.J., Kearsey, M.J. & Davies, P. (1978). The power of the classical twin study. Heredity, **40**, 97-116.

Mascie-Taylor, C.G.N. (1977). Migration and gene flow in Drosophila and man. Unpublished Ph.D. thesis, University of Cambridge, England.

Mascie-Taylor, C.G.N. (1986). Assortative mating and differential fertility. Biology & Society, **3**, 169-170.

Mascie-Taylor, C.G.N. (1988). Assortative mating in a contemporary British population. Annals of Human Biology, **14**, 59-68.

Mascie-Taylor, C.G.N. (1988). Spouse similarity and convergence. Behavior Genetics (in press).

Mascie-Taylor, C.G.N. & Boldsen, J.L. (1984). Assortative mating for IQ: a multivariate approach. Journal of Biosocial Science, **16**, 109-117.

Mascie-Taylor, C.G.N. & Boldsen, J.L. (1988). Assortative mating and natural selection. Annals of Human Biology (in press).

Mascie-Taylor, C.G.N. & Gibson, J.B. (1979).   A biological survey of a Cambridge suburb: assortative mating for IQ and personality traits. Annals of Human Biology, **6**, 1-16.

Mascie-Taylor, C.G.N. & Vandenberg, S.G. (1988).   Assortative mating due to propinquity and personal preference.   Behavior Genetics (in press).

Michielutte, R. (1972).   Trends in educational homogamy.   Sociology of Education, **45**, 288-302.

Murstein, B.I. (1976).   Who will Marry Whom?   Theories and Research in Marital Choice.   New York: Springer.

Murstein, B.I. (1980).   Mate selection in the 70's.   Journal of Marriage and the Family, **42**, 777-792.

Olsson, U. (1979).   Maximum likelihood estimation of the polychoric correlation coefficient.   Psychometrika, **44**, 443-460.

Pennock-Roman, M.P., Vandenberg, S.G. & Mascie-Taylor, C.G.N. (1988). Assortative mating for psychometric characters.   Annals of Human Biology (in press).

Price, R.A. & Vandenberg, S.G. (1960).   Spouse similarity in American and Swedish couples.   Behavior Genetics, **10**, 59-71.

Price, R.A., Chen, K.-C., Cavalli-Sforza, L.L. & Feldman, M.W. (1982).   Spouse influence and smoking behaviour.   Social Biology, **28**, 14-29.

Rice, J., Cloninger, C.R. & Reich, T. (1980).   Analysis of behavioral traits in the presence of cultural transmission and assortative mating: applications to IQ and SES.   Behavior Genetics, **10**, 73-92.

Roa, D.C., Morton, N.E. & Cloninger, C.R. (1979).   Path analysis under generalized assortative mating. I. Theory.   Genetical Research, **33**, 187-198.

Roberts, D.F. (1977).   Assortative mating in man: husband and wife correlations in physical characteristics.   Bulletin of the Eugenics Society, Suppl. 2, 1-45.

Rockwell, R.E.C. (1976).   Historical trends and variation in educational homogamy.   Journal of Marriage and the Family, **38**, 83-95.

Slatis, H.M. & Hoene, R.E. (1961).   The effect of consanguinity on the distribution of continuously variable characters.   American Journal of Human Genetics, **13**, 28-31.

Spuhler, J.N. (1962).   Empirical studies on quantitative human genetics, In: J. N. Spuhler (ed.),   Genetic Diversity and Human Behavior.   Chicago: Aldine.

Spuhler, J.N. (1968).   Assortative mating with respect to physical characteristics.   Eugenics Quarterly, **15**, 128-140.

Stern, C. (1973).   Principles of Human Genetics,   3rd edn.   San Francisco: Freeman.

Thelen, T.H. (1984).   Minority type human mate preference.   Social Biology, **30**, 162-180.

Thiessen, D. & Gregg, B. (1980).   Human assortative mating and genetic equilibrium: an evolutionary perspective.   Ethology & Sociobiology, **1**.

Vandenberg, S.G. (1972).   Assortative mating, or who marries whom.   Behavior Genetics, **2**, 127-157.

Vernon, P.E. (1979).   Intelligence: Heredity and Environment.   San Francisco: Freeman.

Warren, B.D. (1966).   A multiple variable approach to the assortative mating phenomenon.   Eugenics Quarterly,   **13**,   285-290.

Watkins, M.P. & Meredith, W. (1981).   Spouse similarity in newlyweds with respect to specific cognitive abilities, socioeconomic status, and education. Behavior Genetics, **11**, 1-21.

Zonderman, A.B., Vandenberg, S.G., Spuhler, K.P. & Fein, P.R. (1977). Assortative marriage of cognitive abilities.   Behavior Genetics, **7**, 261-271.

ASSORTATIVE MATING FOR ANTHROPOMETRIC CHARACTERS

C. SUSANNE and Y. LEPAGE

Department of Anthropology, Free University of Brussels, Brussels, Belgium

INTRODUCTION

The term "assortative mating" is used by geneticists to describe non-random mating where phenotype influences the probability of mating between individuals. In most situations like attracts like, and this is called "positive assortative mating". Similarity for a trait is also called "homogamy", and mating within a group is called "endogamy".

Some similarity at the genetic level can theoretically be shown between assortative mating and inbreeding, although assortative mating may be specific for a given trait and involves phenotypes who may have different genotypes. The observed phenotypic correlation (r) is related to the genetic correlation m, as follows:

$$m = rh^2$$

where h^2 is the coefficient of heritability of the phenotype under study.

In studies of inheritance, the degree of assortative mating must be taken into account. Assortative mating increases the variance of a trait: in certain equilibrium conditions (Crow & Kimura, 1970) $V = \frac{V_0}{1-r}$, this means for a correlation between spouses of .25 there would be an increase of variance of 1/3. Estimates of heritability are also influenced by assortative mating, the within-family variance being reduced, and the between-family variance being increased (Susanne, 1975, 1977a, 1984).

Assortative mating produces an increase in the average homozygosity in the population, though to a lesser extent than inbreeding. However, traits for which assortative mating exists are moderately genetically determined, so that the evolutionary effects are also moderate.

In human groups, it is known that assortative mating is a widespread phenomenon for many quantitative traits. Assortative mating in physical traits has been reviewed by Susanne (1967), Spuhler (1968) and Roberts (1977).

DIFFICULTIES OF INTERPRETATION

Studies using anthropometric characters have tended to concentrate on height and weight. For other characteristics the number of published studies is small and the measurements are not always comparable. The homogeneity of the studied populations is very different: the correlations will differ as a result of differences in within-group variance and covariance.

Biological and cultural factors are interrelated in the study of population structure. Barriers to gene flow create subdivided populations. These barriers may be geographical, social and/or cultural in nature. Correlations have been shown between geographical distance and anthropometric differentiation; they suggest that a spatial model of gene flow would explain the observed variation among subpopulations (Relethford et al, 1980).

Age, social class and geographical distances can affect the correlations of assortative mating. For some characteristics the observed correlations between partners are influenced by the common marital life, as the result of common nutrition and life styles. Geographical, educational, social and cultural propinquity, as well as similarity of age and life styles, will influence mate selection and the level of assortative mating. It is the "passive" influence defined by Mascie-Taylor and Vandenburg (1988) which is to be distinguished from the "active" role of personal preference. Although there are difficulties in interpretation, it is possible to make certain generalisations.

GENERAL RESULTS

In most Western populations, positive assortative mating is commonly observed for many anthropometric characters. It means that there is a tendency for spouses to resemble each other more than would be expected by chance. Many psychological and social traits are positively correlated between spouses (Vandenberg, 1972; Coleman, 1977). Social class and educational level also play a key role in mate selection (Eckland, 1968).

Age

In studies of assortative mating, one of the most consistent findings is the high degree of correlation for age; coefficients ranging from 0.51 to 0.99 are observed with a great majority around 0.8 and 0.9.

The age difference between spouses is low but consistent: an average of 2.7 years for the United States (Rele, 1965). As assortative mating is always very significant for age, the interpretation seems evident. However, for a

Table 1:   Assortative mating by age in a rural Zapotec community (Oaxaca, Mexico).   Analysis of the two age groups of the spouses (Malina et al, 1983).

	Husband's age			Wife's age	
18 - <30y		>30y - 74y		15 - <30y		>30y - 73y

r	P	r	P	r	P	r	P	r	P
.96	.001	.60	.001	.92	.001	.61	.001	.89	.001

detailed analysis it would be necessary to take into account the age range used in the samples by the different authors.   Table 1 shows the influence of age range in the variation of the correlation of age between spouses in a rural Zapotec community: lower correlations are observed in the lower age group where the age range is the lowest.   This age pattern can influence the value of some correlations of anthropometric taits when these traits change with age (Susanne, 1967).   The possibility of spurious age effects has to be taken into account.

Measurements such as weight, skinfold thickness, bicristal breadth, circumferences of arm and many facial characteristics such as bizygomatic and bigonial breadths, nose and lip height, and ear dimensions, are not stable during adulthood (Susanne, 1977b).   Similarity of age in a sample with large variations of age of the partners will artificially increase the assortative mating of these traits.   Since age is highly correlated, controlling for the age of husband and wife with partial correlations facilitates the interpretation of the phenotypic assortative mating.

The observed changes in the level of correlations between spouses are, however, not always related to changes associated with ageing, as the results of Malina et al (1983) seem to indicate (Table 2).   In this study of a Zapotec sample, the authors observed a decrease of the assortative mating of stature after controlling for age; dividing the sample into younger and older subsamples resulted in lower correlations in the older group.   As the authors indicated, this increase from the younger to the older group could be related to sex differences in the ageing of stature, but it is also possible that the decision of the parents in a choice of a partner would have been more strictly followed in the older group than in more recent marriages.

The greater endogamy of the older group is perhaps also the explanation of a lower assortative mating with respect to stature (0.07, NS, in the grandparental generation) than in a more exogamous situation (in the same Polish population, 0.20, p <.01, in the parent generation) (Wolanski, 1980), or

Table 2:   Assortative mating in a rural Zapotec community (Malina et al, 1983)

| | | | | | Husband's age | | | | Wife's age | | | |
| | | | | | <30y | | >30y | | <30y | | >30y | |
	r	P	r*	P	r	P	r	P	r	P	r	P
stature	.35	.001	.24	.027	.34	.009	.02	NS	.31	.012	.18	NS
weight	.01	NS	.01	NS	-.05	NS	.10	NS	-.07	NS	.13	NS
triceps skinfold	.16	NS	.15	NS	-.11	NS	.51	.011	-.04	NS	.56	.012

* partial correlation (controlled for ages of both spouses)

Table 3:   Assortative mating for physical traits

	r	$r_{controlled}$		r	$r_{controlled}$
height	.178	–	head length	.045	.020
weight	.054	-.023	head breadth	.133	.140
weight at marriage	.070	-.074	frontal breadth	.183	.218
sitting height	.110	.084	bizygomatic breadth	.055	.102
arm length	.111	.079	bigonial breadth	.102	.118
biacromial breadth	.042	.003	chin-nasion	.207	.183
wrist circumference	-.085	-.078	nose height	.208	.271
ankle circumference	.122	.094	lip height	.155	.148
arm skinfold thickness	.328	.332	physiognomic height	.060	.042

$r_{controlled}$: partial correlation taking height into account.

in the rural Oxfordshire villages (Harrison et al, 1976) (-0.021 when married before 1940, 0.088 in the period 1940-50, 0.339 in the period 1950-60, and 0.242 after 1960).

On the other hand, some correlations increase from the younger to the older subsamples: in the case of fatness, expressed by the triceps skinfold, and perhaps also weight, we are observing the results of communal living, of the similar habits of food intake and of way of life. This is the case in the results of Table 2 and in the Polish situation (Wolanski, 1980; 0.05 in the parent generation and 0.41 in the grandparent generation for the skinfold thickness).

Height

For populations of European origin, the median spouse correlation of height is about 0.2. This pattern could also account for other observed "secondary" assortative mating, since almost all physical traits are correlated with stature (Table 3). Controlling for height of both partners results in a lowering of some correlations between spouses. It is specially evident for weight, for sitting height, length of arm, biacromial breadth, head length, but of course less for measurements where the correlation with height is low such as for skinfold thickness or facial traits.

Moreover, Plomin et al (1977) observed that the assortative mating observed for unwed biological parents of children placed for adoption was as high as for married couples; only for education and social background was the resemblance less evident.

The positive assortative mating of height is partly the result of the social expectation that the male should be taller (Gillis et al, 1980). This is a cultural influence which can differ between societies. It is perhaps also influenced by the social mobility. The inter-generational social mobility shows clearly that upward mobility is related to higher stature, and downward mobility to lower stature (Mascie-Taylor, 1984).

Positive assortative mating does not exclude some non-linearity. Indeed, McManus and Mascie-Taylor (1984) showed that in the case of women of tall stature (>1.80 metres) a negative assortative mating is observed. With taller women, there is a lack of appropriate partners (taller or even of equal height), which results in a lower assortative mating.

In any case, a generalisation emerges: spouse correlations for stature are consistently found and these are moderate: these remain even after controlling for age of the spouses (Susanne, 1967; Johnston, 1970; Malina et al, 1983). It is a confirmation of the classic study of Pearson and Lee (1902) where the assortative mating for stature was 0.28.

Table 4 gives a list of correlations for samples of European and non-European origin for stature. It is difficult to interpret these correlations as measures of assortative mating *per se* since many factors may be involved. We will analyse some of these factors, such as residential propinquity and ethnic variation. Moreover, the samples may differ in within-group variances, in the homogeneity of socioeconomic sampling and in age. It is difficult to judge from most studies if the observed correlations between husbands and wives are really representative of assortative mating or if they are the indirect result of the great heterogeneity of the samples, or even the influence of morphological changes after marriage resulting from common patterns of nutrition, life style and physical exercise. It is in these conditions not surprising to find great variability in the observed correlations. Even in the same population, but in different socioeconomic groups, the level of correlation between spouses may be

Table 4   Assortative mating for stature

		r	p
Population of the European origin			
Infertile couples (S)	Pomerat (1936)	.63	<.01
German (Breslau)	Schwidetzky (1941)	.44	<.01
American	Smith (1946)	.38	<.01
American	Garn et al (1979)	.37	<.01
American	Schiller (1932)	.36	<.05
American	Mueller et al (1976)	.34	<.01
English	Huntley (1967)	.34	<.01
New York	Davenport (1917)	.33	<.01
Scottish (migrants US)	Willoughby (1933)	.32	<.01
American	Burgess et al (1944)	.31	<.01
Emilian (Italy)	Graffi-Benassi (1936)	.305	<.01
English	Pearson (1899)	.29	<.01
Ann Arbor (USA)	Schwidetzky (1941)	.29	<.01
English	Pearson et al (1902)	.28	<.01
Polish	Wolanski et al (1970)	.28	<.01
Belgian	Susanne (1967)	.277	<.01
UK	Mascie-Taylor (1985)	.277	<.01
Bulgarian	Nicolaeff (1931)	.273	<.01
American (upper SES)	Malina (1986)	.27	<.05
Finkenwärder	Scheidt (1930)	.27	NS
Polish (Silesian coal)	Siniarska (1984)	.259	<.05
German (Eugenfeld)	Nicolaeff (1931)	.254	<.05
English (Oxfordshire)	Harrison et al (1976)	.232	<.05
Polish (Belchatow)	Siniarska (1984)	.231	NS
Ukrainian Jews	Nicolaeff (1931)	.226	<.01
Lapps (Sweden)	Elston (1961)	.215	<.01
Sardinian	Tomici (1939)	.204	<.01
American	Garn et al (1979)	.20	<.01
American	Spuhler (1968)	.20	<.01
Polish	Wolanski (1980)	.20	<.01
Polish	Wolanski (1973)	.199	<.01
Hungarian (migrants US)	Willoughby (1933)	.19	NS
English	Pearson et al (1902)	.18	<.01
Belgian	Van Scharen et al (1973)	.178	NS
English	Pearson (1899)	.178	<.01
Hungarian	Ugge (1931)	.174	NS
Italian (migrants US)	Willoughby (1933)	.17	<.05
Italian (Southern)	Ugge (1931)	.169	<.01
Hebrew	Ugge (1931)	.15	<.01
Greek	Nicolaeff (1931)	.141	NS
Sicilian	Genna (1941)	.139	<.05
Polish	Rosinski (1923)	.138	<.01
Polish	Ugge (1931)	.132	NS
Scottish	Ugge (1931)	.124	NS
Hebrew (migrants US)	Willoughby (1935)	.12	<.05
Sicilian	Ugge (1931)	.118	<.05
Sicilian (migrants US)	Willoughby (1933)	.11	NS
Polish (Lublin coal)	Siniarska (1984)	.101	NS
Bohemian (migrants US)	Willoughby (1933)	.10	NS
Ukrainian	Nicolaeff (1971)	.094	NS

Table 4, contd.

English	Pearson (1899)	.093	NS
Polish	Wolanski (1980)	.07	NS
Russian	Nicolaeff (1931)	.068	NS
Polish (Suwalki)	Simiarska (1984)	.064	NS
Bohemian	Ugge (1931)	.057	NS
American (low SES)	Malina (1986)	.03	NS

Non-European origin

Rehoboth	Scheidt (1930)	.52	<.05
Zapotec (Mexico)	Malina et al (1983)	.35	<.01
Cashinahua (Peru)	Johnston (1970)	.346	<.01
Chad (monogamous)	Crognier (1977)	.31	<.01
Mexican (US)	Malina (1986)	.25	<.05
Mestizo (Colombia)	Mueller (1975)	.21	<.01
Chad (poligamous)	Crognier (1977)	.21	<.01
Mexican (US)	Malina (1986)	.15	NS
Mandinka (Gambia)	Roberts (1977)	.146	NS
Nasioi (Solomon Is.)	Baldwin et al (1973)	.12	NS
Mexican (US)	Malina (1986)	.12	NS
Baegu (Solomon Is.)	Baldwin et al (1973)	.12	NS
Calapos (Brazil)	Da Rocha (1971)	.08	NS
Japanese	Furusho (1961)	.072	NS
Negroes (American)	Mueller et al (1976)	.06	NS
Caingang (Brazil)	Da Rocha (1971)	.06	NS
Lau (Solomon Is.)	Baldwin et al (1973)	-.05	NS
Kwaio (Solomon Is.)	Baldwin et al (1973)	-.10	NS
Ramah Navaho	Spuhler (1968)	-.18	NS
Seminole (US)	Pollitzer et al (1970)	-.26	NS
Xavante (Brazil)	Da Rocha (1971)	-.32	NS

different. At least for stature, the correlation may be higher in the upper socio-economic status translating perhaps more factors of personal choice and a greater exogamy than in lower social groups where the social system will be more limiting and where the group will be more highly endogamous.

Similar results have been observed by Siniarska (1984) in the Polish population: the coefficient of correlation between spouses is higher in an industrialised area with greater exogamy than in a rural area with more endogamy. Similar results, however, were not obtained by Harrison et al (1976), who found that in rural Oxfordshire villages the highest correlations were observed in social class 3, or by Mascie-Taylor (1987) where, in a UK sample, the highest values were observed in classes 1 and 5.

The causes of assortative mating remain obscure: direct personal choices and indirect influences of social and geographic factors are involved, as well as the size of the population with which rate of endogamy will be related.

The data for non-European samples are more limited. Examples are:

Japanese (Furusho, 1961), American blacks (Mueller et al, 1976) in industrial countries, and, for rural samples, Ramah Navaho (Spuhler, 1968), Cashinahua from Peru (Johnston, 1970), Solomon Islanders (Baldwin et al, 1973), Colombian Mestizo (Mueller, 1975), and Zapotec Indians in the valley of Oaxaca (Malina et al, 1983).

The results can, in the case of more traditional communities, be influenced by many other factors than in industrial or urban conditions. Indeed, these communities are perhaps more endogamous; life style and nutritional conditions may interact in the husband-wife correlations; ageing effects can show patterns different from the industrial countries; and moreover the system of marriage can be very different. In many cases parents will choose the partner. Phenotypic assortative mating, as expressed in husband-wife correlations, cannot in these situations be interpreted as a function of the partners' personal factors of choice but rather as a function of a social system.

Table 4 shows that assortative mating is sometimes negative in some non-European populations: the mean value is lower than in European populations. Absence of personal choice, endogamy, ethnic basis in determining the ideal marriage are possible explanations of this variability in assortative mating.

It is possible that assortative mating for stature would increase relatively to random mating. The relationship between population size and correlation between spouses described by Wolanski and Siniarska (1984) could be an illustration of this effect: lower coefficients in smaller population size with potentially high endogamy, maximal mean values of .25 in larger exogamous populations. We will see that some differences observed in terms of the age of spouses are perhaps also related to differences in degree of random mating.

Weight

Assortative mating for weight is significantly lower than that for height. The correlation between spouses is more variable than for stature: some correlations are even higher than for stature but very often the correlations are low and not significant. More than for stature, we must take into account environmental conditions in the interpretation of these results, ageing effects and conjugal living for older spouses, assortative mating, but also number of pregnancies must be taken into account in younger spouses.

Other traits

Extensive reviews of assortative mating with respect to anthropometric measurements have been published by Susanne (1967), Spuhler (1968) and Roberts (1977).

Longitudinal body measurements are also positively correlated between partners. Positive assortative mating with respect to some breadth measurements and circumferences is also observed, such as for bicristal breadth, waist and arm circumferences, and for skinfold thicknesses. Significant assortative mating for head and face measurements is mostly positive, such as for head breadth, frontal bizygomatic, and bigonial breadth, facial height, nose and lip height.

Discussion

It is possible that when assortative mating is studied using data based on marriages of short duration the degree of assortative mating is underestimated: the unstable marriages could have a lower degree of similarity. Cattell et al (1967) found that stable couples showed a higher assortative mating for personality and intelligence than in unstable marriages. Kiser (1968) showed a higher fertility of the couples who are homogamous for education. Dean et al (1978) observed a low homogamy for age and education in a sample of divorced women. Bentler et al (1978) in a comparison of divorced couples with couples still married after four years showed less homogamy in the divorced sample.

It is also possible that for other traits the similarity between spouses increases after marriage as a consequence of common environmental influences. However, Harrison et al (1976) failed to observe an influence of the length of marriage on assortative mating of IQ, of personality traits or of anthropometric traits, even for weight which is environmentally labile.

CAUSES OF THE SIMILARITIES

The causes remain obscure: direct choice in terms of physical attractiveness, indirect association with residential propinquity as well as length of time between first meeting and marriage, length of marriage or even common schooling are all factors that can influence the similarity between spouses.

Physical attractiveness

Physical attractiveness of spouses and of engaged couples is moderately correlated. In a study of Murstein (1976) the correlation between ratings of photographs by eight judges of physical attractiveness was .38. Very young children are already aware of cultural prescriptions about attractiveness. But physical attraction has indirectly powerful effects (Bar-Tal et al, 1976; Krebs & Adinolfi, 1975): beautiful persons are perceived as being more successful, confident, sincere, and having a more prestigious social level.

This halo effect could influence indirectly the relationship between (female) attractiveness and social mobility. In fact, female physical attractiveness is more related to mobility through marriage of women of working-class origin, education is more related to social mobility for women of middle-class origin (Elder, 1969; Taylor et al, 1976). These authors found that attractiveness is correlated with husband's social level at .369 for women whose fathers were workers, but only .148 for farmers' daughters, similar lower correlations being observed for clerical/craftsmen classes.

Correlations for physical measurements may be an indirect influence of assortative mating for attractiveness: this could be the case for tallness and slimness in our European standards. Interactions with social class, education, intelligence and occupation are surely present at this level.

Residential propinquity

Studies of neighbourhood knowledge (Boyce et al, 1967) showed clearly that people tend to marry those who live nearby. The marital distances depend on the social level of the samples (Harrison et al, 1971; Coleman, 1973; Susanne, 1982). In any case, isolation by geographical origin remains very efficient (Lepage, 1979; Coleman, 1984).

Race also strongly limits marital choice. In Hawaii, the rate of intermarriage has been estimated to be 12.1% (Vandenberg, 1972). But in the United States, even now, frequency of black-white intermarriages is estimated to be less than 1% (Kephart, 1977). The ethnic barrier is great when the morphological variation between populations is large (Morton et al, 1967). Endogamy for religious group is estimated by Kephart (1977) to be 65 to 85%. This barrier is thus also rather rigid (Eckland, 1968).

Socioeconomic factors also remain very important (Schwidetsky, 1950; Segalen et al, 1971; Price & Vandenberg, 1979). This social homogamy appears even in a comparison of the professions of the respective fathers of both partners (±50% of concordance, Susanne & Van Scharen, 1973, 1974; Van Scharen & Susanne, 1974). An important factor in marital choice is also educational class (Kiser, 1968; Garrison et al, 1968; Johnson et al, 1976; Williams, 1975).

Anthropometric characters vary as a result of race, social or educational status, or even because of geographical isolation, and the correlation between partners will increase "artificially" when the sample is heterogeneous for one of these factors.

Mating group

The importance of a trait in terms of mate choice will vary between different social groups and between generations, with current new generations perhaps less influenced by social status than previous ones. Urban and rural contrasts also exist. In France, marital geographic distances are greater in urban areas than in rural areas (Girard, 1964). The circumstances of mating show substantial differences. In rural countries the mating group includes almost all individuals of the village, the mating group is geographically rather well defined. Social homogamy, although less important than in urban countries, varies according to social group: the potentiality to find a partner in the village is a function of socioeconomic factors.

In towns, the mating group of one individual will differ from the mating group of his neighbour. Many factors will have a greater influence in towns than in rural countries: ethnic, racial, social, religious, political, psychological factors - all contribute to the diversity and heterogeneity of the mating groups existing in towns. The difference between mating groups of towns and rural countries is that these groups are geographically enlarged in towns while being more heterogeneous and continuously changing as a result of geographical and social migration.

The variation in anthropometric characters has been shown to be correlated with geographic distance (Relethford et al, 1980), so that the existence of a mating group will possibly result in assortative mating.

Ethnic variation

Except in societies with arranged marriages, mate selection can be influenced by different ethnic factors. Physical attractiveness, homogamy and endogamy, barriers vary in each population. Correlation between spouses integrates all these factors. It is perhaps not surprising to observe in these circumstances that at least for height (Figure 1) the correlations observed in non-industrialised countries of non-European origin are overall lower than in industrialised countries. This difference is less evident for weight. For the other measurements, the low number of published results makes a comparison impossible (length of head and bizygomatic breadth, for instance, Figure 1).

DISCUSSION

Thus, there are prescriptions (male taller and older, but not too much older) and there are norms at racial and social level that persons have acquired and follow almost automatically. In a homogeneous group, propinquity will

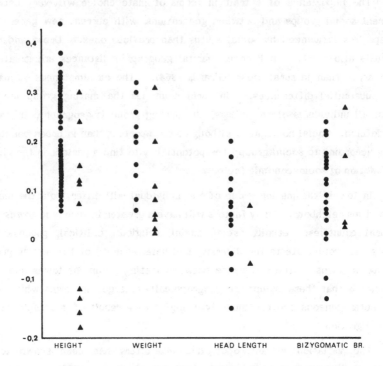

Figure 1:   Comparison of correlations of assortative mating concerning
                different samples of European origin and of non-European origin.

determine mate selection;   in a socially mixed group norms will be more
powerful than propinquity.   When social or psychological assortative mating is
observed, the interpretation is mostly geographical propinquity, social similarity
and personal selection.   Social and geographic distances act as barriers.   These
barriers probably also influence physical assortative mating;   individuals of the
same social group and same geographical area are more likely to have similar
anthropometric traits than would two randomly chosen persons.   Assortative
mating can have this global indirect origin.

        For some physical traits, a more active selection of spouses may occur
through personal preferences.   We can imagine that assortative mating of
fingerprints would be influenced by geographical barriers, of personality traits by
personal selection, and of facial traits probably by propinquity and selection.
Even in western countries, ethnic differences in personal preferences exist.
Hansen (1977) reports differences in criteria used in the choice of a mate
between white and black students: white students tended to emphasise traits

such as "is intelligent", "has good sense"; black chose external traits such as "is appropriately dressed", although these differences are perhaps of socioeconomic origin.

Each population is characterised by barriers of different origin and of different intensity. A mating group corresponds to specific geographical social, ethnic, religious and political barriers, and sometimes occupation, education, age and civil status. The "choice" of partner is thus limited, the way future spouses meet is characteristic of what is "allowed" by the population to which the individuals belong. Marriages in western countries are not arranged, such as is the case in caste societies (Sanghvi, 1970). But, even in western countries, marital choice is influenced by parents. Free choice is apparent, spouses are chosen mostly from a group of "eligibles". The consequence of these choices is very likely an assortative mating of anthropometric characters.

Mascie-Taylor et al (1988) tried to distinguish the general "passive" selection from a more active personal choice within the range of the possible partners. They studied an urban sample for IQ test (using the Wechsler scale) and personality (Eysenck inventory). Using partial correlation coefficient analysis to allow for social, educational, geographical and familial propinquity and stepwise multiple regression, they dissociated the passive and active components of choice. For IQ the original explained variance (R^2) was 14%, representing about 10% related to propinquity and only 4% to an active choice. For personality such as extraversion/introversion, the passive and active contributions were almost equal. Mascie-Taylor (1987) applied the same approach to height and weight of the parents of children in the National Child Development Study: in a multiple regression the effects of social class, region, years of education and age on the observed assortative mating were very small: assortative mating for height declined only by 0.015 and weight by zero.

The factors of mate selection being variable according to the studied population, ethnic variation of assortative mating can result. The ideal marriage is ethnically variable, but this ideal will be probably more strictly followed in isolated populations. This variablity of the observed coefficients of correlation has been observed by Crognier (1973) in a comparison of monogamous or polygamous marriages in Chad. In industrial European societies, the importance each individual gives to psychological, social, physical and aesthetic factors in the choice of a partner is highly variable. If assortative mating of physical characteristics is not yet elucidated, it is because it is an integration of these different factors and barriers.

REFERENCES

Baldwin, J.C. & Damon, A. (1973).   Some genetic traits in Solomon Island populations. American Journal of Physical Anthropology, **39**: 195-202.

Bar-Tal & Saxe, L. (1976).   Perceptions of similarly and dissimilarly attractive couples and individuals. J.Personal.Soc.Psychol., **33**: 772-781.

Bentler, P.M. & Newcomb, M.D. (1978).   Longitudinal study of marital success and failure. J.Consult.Clin.Psychol. **46**: 1053-1070.

Boyce, A.J., Küchemann, C.H. & Harrison, G.A. (1967).   Neighbourhood knowledge and the distribution of marriage distance.   Ann.Hum.Genet., **30**: 335-338.

Burgess, E.W. & Wallin, P. (1944)   Homogamy in personality characteristics. J.Abn.med.Soc.Psychol., **39**, 475-481.

Cattell, R.B. & Nessebroad, J.R. (1967).   Likeness and completeness theories examined by sixteen personality factors measured on stably and unstably married couples.   J.Pers.Soc.Psychol. **7**: 351-361.

Coleman, D.A. (1973).   Marriage movement in British cities,   In: D. F. Roberts & E. Sunderland (eds.),   Genetic Variation in Britain, pp. 33-57.   London: Taylor & Francis.

Coleman, D.A. (1977).   Assortative mating in Britain,   In: R. Chester & J. Peel (eds.), Equalities and Inequalities in Family Life, pp. 17-52.   Academic Press.

Coleman, D.A. (1984).   Marital choice and geographical mobility,   In: A. J. Boyce (ed.),   Migration and Mobility: Biosocial Aspects of Human Movement, pp. 19-55. London: Taylor & Francis.

Crognier, E. (1973).   Les adaptations morphologiques d'une population africaine au biotope tropical: les Sara du Tchad. Bull.Soc.Anthrop.Paris, **10**: 3-152.

Crognier, E. (1977).   Assortative mating for physical features in an African population from Chad. J.Hum.Evol., **6**: 105-114.

Crow, J.F. & Kimura, M. (1970).   An Introduction to Population Genetics Theory. New York: Harper & Row.

da Rocha, F.J. (1971).   Antropometria en Indigenas Brasileiros.   Ministerio da Educacao, Porto Alegre.

Davenport, C.B. (1917).   Inheritance of stature. Genetics, **2**: 327-329.

Dean, G. & D. Gurak (1978).   Marital homogamy the second time around.   J. Marr. & Family, **40**: 559-570.

Eckland, B.K. (1968).   Theories of mate selection. Eugen.Quart., **15**: 71-84.

Elder, G.H. (1969).   Appearance and education in marriage mobility. Amer.Sociol.Rev. **34**: 519-533.

Elston, R.M. (1961).   Some data on assortative mating in man. Proceedings of the Second International Congress of Human Genetics, **3**: 2005-2006.

Furusho, T. (1961).   Genetic study on stature. Jap.J.Hum.Genet., **6**: 78-101.

Garn, S.M., Cole, P.E. & Bailey, S.K. (1979).   Living together as a factor in family-line resemblances. Human Biology, **51**(4): 565-587.

Garrison, G.H., Anderson, V.E. & Reed, S.C. (1968).   Assortative mating. Eugen.Quart. **15**: 133.

Genna, G. (1941). Aspetti antropologici dell'assortimento matrimoniale. Archivio per l'Antropologia e la Etnologia, **71:** 5-25.

Gillis, J.S. & Avis, W.E. (1980). The male taller norm in mate selection. Pers.Soc.Psychol.Bull. **6:** 396-401.

Girard, A. (1964). Le Choix du Conjoint. Paris: Presses Universitaires.

Graffi-Benassi, E. (1936). Contributo allo studio delle rassomiglianze fra coniugi. Rivista di Antropologia, **31.**

Hansen, S.L. (1977). Dating choices of high school students. The family coordinator, **26:** 133-138.

Harrison, G.A., Gibson, J.B. & Hiorns, R.W. (1976). Assortative mating for psychometric, personality and anthropometric variation in a group of Oxfordshire villages. J.Biosoc.Sci. **8:** 145-153.

Harrison, G.A., Hiorns, R.W. & Küchemann, C.F. (1971). Social class and marriage patterns in some Oxfordshire populations. J.Biosoc.Sci., **3:** 1-12.

Huntley, R.M.C. (1967). Some problems in the study of quantitative variation in man. In: S. G. Pickett (ed.), Endocrine Genetics, pp. 229-248. Cambridge: Cambridge University Press.

Johnson, R.C., Defries, J.C., Wilsin, J.R., McClean, G.E., Vandenberg, S.G., Ashton, G.C., Mi, M.P. & Rashad, M.N. (1976). Assortative mating for specific cognitive abilities in two ethnic groups. Hum.Biol. **48:** 343-352.

Johnston, F.E. (1970). Phenotypical assortative mating among the Peruvian Cashinahua. Social Biology, **17:** 37-42.

Kephart, W.M. (1977). Family, Society and the Individual. Boston: Houghton-Mifflin.

Kiser, C.V. (1968). Assortative mating by educational attainment in relation to fertility. Eugen.Quart., **15:** 98-112.

Krebs, D. & Adinolfi, A.A. (1975). Physical attractiveness, social relations and personality style. J.Personal.Soc.Psychol., **31:** 245-253.

Lepage, Y. (1979). Cent-vingt annees de choix du conjoint a Alle-sur-Semois. Population, **6:** 1152-1161.

McManus, I.C. & Mascie-Taylor, C.G.N. (1984). Human assortative mating for height: non-linearity and heteroscedasticity. Hum.Biol., **56:** 617-623.

Malina, R.M. (1986). Personal communication.

Malina, R.M., Selby, H.A., Buschang, P.H., Aronson, W.L. & Little, B.B. (1983). Assortative mating for phenotypic characteristics in a Zapotec community in Oaxaca, Mexico. J.Biosoc.Sci., **15:** 273-280.

Mascie-Taylor, C.G.N. (1984). The interaction between geographical and social mobility. In: A. J. Boyce (ed.), Migration and Mobility: Biosocial Aspects of Human Movement, pp. 161-178. London: Taylor & Francis.

Mascie-Taylor, C.G.N. (1987). Assortative mating in a contemporary British population. Ann.Hum.Biol. **14,** 59-68.

Mascie-Taylor, C.G.N. & Vandenberg, S.G. (1988). Assortative mating due to propinquity and personal preference. Behavior Genetics (in press).

Morton, N.E., Chin, S.S. & Ming, P.M. (1967). Genetics of Inter-racial Crosses in Hawaii. Basel: Karger.

Mueller, W.H. (1975). Parent-Child and Sibling Correlations and Heritability of Body Measurements in a Rural Colombian Population. Ph.D. thesis. Austin: University of Texas.

Mueller, W.H. & Malina, R.M. (1976). Differential contribution of stature phenotypes to assortative mating in parents of Philadelphia black and white schoolchildren. Amer.J.Phys.Anthrop., 45(2): 269-275.

Murstein, B.I. (1976). Who will Marry Whom? Theories and Research in Marital Choice. New York: Springer.

Nicolaeff, L. (1931). Les correlations entre les caracteres morphologiques des epoux. Anthropologie, 41: 75-93.

Pearson, K. (1899). Data for the problem of evolution in man, III. Proc.Roy.Soc., 66: 23-32.

Pearson, K. & Lee, A. (1902). On the laws of inheritance in man: I. Inheritance of physical characters. Biometrika, 2: 357-462.

Plomin, R., Defries, J.C. & Roberts, M.K. (1977). Assortative mating by unwed biological parents of adopted children. Science, 196: 449-450.

Pollitzer, W.S.D., Rucknagel, D., Tashian, R., Shreffler, D.C., Leyshon, W.C., Namboodiri, K. & Elston, R.C. (1970). The Seminole Indians of Florida: morphology and serology. Amer.J.Phys.Anthrop., 32: 65-82.

Pomerat, C.M. (1936). Homogamy and infertility. Human Biology, 8: 19-22.

Price, R.A. & Vandenberg, S. (1979). Spouse similarity in American and Swedish couples. Behav.Genet., 10: 59-71.

Rele, J.R. (1965). Trends and differentials in the American age at marriage. Milbank Memorial Fund Quarterly, 43: 224.

Relethford, J.H., Lees, F.C. & Crawford, M.H. (1980). Population structure and anthropometric variation in rural Western Ireland: migration and biological differentiation. Ann.Hum.Biol., 7: 411-428.

Roberts, D.F. (1977). Assortative mating in man: husband/wife correlations in physical characteristics. Suppl.Bull.Eugen.Soc., 2: 1-45.

Rosinski, B. (1923). Charakterystyka antropologiczna ludnosci pow. Pultuskiego, "Komos", 48.

Sanghvi, L.D. (1970). Changing patterns of caste in India. Soc.Biol., 17: 299-301.

Scheidt, W. (1930). Annahme und Nachweis von Rassenvermischung. Zeitschrift für Morphologie und Anthropologie, 27: 94-116.

Schiller, B. (1932). A quantitative analysis of marriage selection in a small group. J.Soc.Psychol., 3: 297-314.

Schwidetsky, I. (1941). Standes und Berufstypus in Breslau. Zeitschrift fur Rassenkunde, 12: 351-379.

Schwidetsky, I. (1950). Grundzüge der Völkerbiologie. F. Enke Verlag, Stuttgart. Biologie der Partnerwahl, 184-206.

Segalen, M. & Jacquard, A. (1971). Choice du conjoint et homogamie. Population, 26: 487-498.

Siniarska, A. (1984). Assortative mating of parents and sib-sib similarities in offspring. Studies in Human Ecology, 5: 95-112.

Smith, M. (1946). A research note on homogamy of marriage partners in selected physical characteristics. Amer.Sociol.Rev., 11: 226-228.

Spuhler, J.N. (1968). Assortative mating with respect to physical characteristics. Eugen.Quart., **15**: 128-140.

Susanne, C. (1967). Contribution a l'etude de l'assortiment matrimonial dans un echantillon de la population belge. Bull.Soc.roy.belge Anthrop.Prehist., **78**: 147-196.

Susanne, C. (1975). Genetic and environmental influences on morphological characteristics. Ann.Hum.Biol., **2**: 279-287.

Susanne, C. (1977a). Heritability of morphological characters. Hum.Biol., **49**: 573-580.

Susanne, C. (1977b). Individual age changes of the morphological characteristics. J.Hum.Evol., **6**: 181-189.

Susanne, C. (1982). Biodemographical structure of the Belgian population. In: N. Wolanski & A. Sinarska (eds.), Ecology of Human Populationis, pp. 269-296. Ossolineum Wroclaw.

Susanne, C. (1984). Methods in human growth genetics, In: C. Susanne (ed.), Genetical Environmental Factors during the Growth Period, pp. 61-80. Plenum.

Susanne, C. & Van Scharen, F. (1973). Partnerkeuze: biodemografische aspekten. Bevolking en gezin, **3**: 427-444.

Taylor, P.A. & Glenn, N.D. (1976). Utility of education and attractiveness for female status attractiveness through marriage. Amer.Sociol.Rev., **41**: 484-498.

Tomici, L. (1939-1940). Fattori somatici dell'attrazione matrimoniale nei coniugu sassaresi, Genus, **4**: 21-54.

Ugge, A. (1931). Sulla rassomiglianza fra coniugi per alcuni caratteri somatici. Contributi del Laboratorio de Statistica. Pubblicazioni dell'Universita Cattolica del Sacro Cuore, Serie 8, vol. 6, Milano.

Vandenberg, S.G. (1972). Assortative mating, or who marries whom. Behav.Genet., **2**: 127-157.

Van Scharen, F. & Susanne, C. (1974). Assortement matrimonial: etude de quelques caracteres anthropologique et psychologiques. Homo, **25**: 146-158.

Williams, T. (1975). Family resemblance in abilities: the Wechsler scales. Behav.Genet., **5**: 405-409.

Willoughby, R.R. (1933). Somatic homogamy in man. Hum.Biol., **5**: 690-705.

Wolanski, N. (1973). Assortative mating in the Polish rural population. Studies in Human Ecology, **I**: 182-188.

Wolanski, N. (1980). Heterosis and homosis in Man, In: Physical Anthropology of European Populations, pp. 213-233. Hague: Mouton.

Wolanski, N. & Chrastek-Spruch, H. (1970). Wysokosc ciala i wiek rodzicow a diugosc i ciezar ciala noworodkow oraz dynamika rozwoju niemowlat. Przeglad Antropologiczny, **36**(1-2), 53-72.

Wolanski, N. & Siniarska, A. (1984). Species module and assortative mating in man, J.Hum.Evol., **13**: 247-253.

APPLICATION OF
SURNAME FREQUENCY DISTRIBUTIONS
TO STUDIES OF MATING PREFERENCES

G. W. LASKER

Wayne State University, Detroit, Michigan, U.S.A.

INTRODUCTION

The study of surnames may seem far removed from the modern study of genetics with its DNA analyses and gene splicing. Nevertheless, surnames are inherited and can be used in models of genetic structure (Lasker, 1985). There are shortcomings, of course: firstly the degree of departure from fulfilment of the basic assumption that a surname in common means descent in common (see, for instance, Ellis & Starmer, 1978); and secondly it must be remembered that surnames are not expressed biologically and are therefore not subject to biological adaptation in the course of development of individuals or in the population through natural selection. This latter may be an advantage, however, since the lack of biological adaptation in surnames provides a simpler model for comparison with biological characteristics - most of which are subject to some degree of adaptation.

Surnames were invented to identify individuals with their genetic lineages. That is, ever since the time other naming systems were transformed into hereditary afternames, surnames have been used to convey genetic information. Even the use of statistics on surnames for this purpose is now old: George Darwin, the son of Charles Darwin, introduced this use 111 years ago. There are limitations to the uncritical application of surname statistics, however. Common surnames, such as Smith and Jones, contribute most of the information contained in surname statistics, but it is the relatively rare surnames such as Cape, Ayers or Dore that are most likely to link two individuals born with the same one, whereas two individuals named Smith are likely to be descended from two different medieval blacksmiths who were entirely unrelated to each other.

What follows will deal primarily with West European systems of surnames. Other populations present special problems. Most of the populations of anthropological interest use no surnames or have begun to use them only recently. Surnames are of little use to tribal peoples. In so-called primitive societies people choose their mates in terms of kinship (for instance cross-cousins) or in terms of clan or lineage membership. The basis of such marriages in kinship structure has dominated much of social anthropology. That is, the study of the social significance of marriage systems based on kinship has dealt with the non-literate societies which lack vital records and in which genetic definitions of kinship do not necessarily correspond exactly with the terms through which social relationships are governed. Although kin-based marriage is important biologically because marriage to a relative results in inbreeding and increased genetic homozygosity in the offspring, genetic studies of inbreeding have usually best been pursued in literate societies which have a kinship nomenclature with precise genetic meanings and which keep records. These are the very societies which have used surnames for some time, but generally they are societies which pay relatively little attention to kinship in respect to choice of mates.

The motives involved in choice of mates in Western European type societies, with whose surnames we shall be concerned, have little to do with kinship as such. The potential husbands, as judged after the fact from actual pairs of mates, tend to be just a few years older, just a few centimetres taller, of similar economic status, similar educational and intellectual attainment, same religious background, of the same nationality and ethnicity, and more often resident (hence often also born) near rather than far away from their wives. The crucial condition for mating is the opportunity to meet and consciously or subconsciously to consider each other's social, psychological and physical qualities. The relative importance of separate factors varies greatly, and the degree of primacy of each is difficult for the investigator to determine; some factors which may seem important may merely be incidental concomitants of others. In the Western World, as elsewhere, there is an incest taboo against mating with close relatives. Unlike many other societies, however, there is virtually no explicit preference for mating with some class of more distant kin.

Among certain ethnic minorities and more often in the past than now, there has been more frequent marriage with kin than would be expected at random. That is, there is a positive selection for kin as well as the negative selection of the incest taboo. Even in these groups it may be more a matter of how the group is defined than an actual felt preference that leads to these marriages. Selection for similar social and economic characteristics in a small ethnic minority may leave few other than relatives to choose from.

SURNAMES AND GENETIC STRUCTURE OF POPULATIONS

In societies of Western European origin, surnames are generally passed on through the male line and are inherited like alleles of a gene. Except that daughters as well as sons receive their fathers' surnames and that women actually assume their husbands' surnames, surnames behave like alleles of a Y-linked gene. The exceptions are not a serious problem because maiden names can be substituted for married ones, and descent from father to daughter is as interesting biologically as that from father to son.

Although surnames do not reveal motives for selection of mates, the distribution of surnames does reflect the net effect on the population's genetic structure of such selection. The degree of population structure in Western societies shown by surname distributions does not result from kin preferences of the type shown by many non-literate societies, but the extent of a similar result can be measured in the structure implied by surname distributions.

Surname distributions model breeding structure in two time frames: the long span since surnames came into use, and the one-generation span involved in mate selection. The general distribution of frequencies of surnames in a population is present because of all the events affecting distribution since the surnames arose. In that period (in South and West Europe generally 700-900 years) the migrations and differential survival rates have cumulatively led to the present frequencies and geography of surnames among descendants. This is the accumulated aspect of population structure. It is represented by the random component of the inbreeding coefficient (F_r), by the Coefficient of Relationship by Isonymy (R_i) and by the random component of the tendency to repetition of the same pair of surnames in different marriages (RP_r).

The other time frame one can study by surnames is the single generation. Data on the surnames brought together by unions of mates yield information on the dynamics of structure formation - mate selection and inbreeding of the specific generation that yields the surname data. Data on the surnames of married pairs permit calculation of the non-random component of the inbreeding coefficient (F_n) and of the non-random residual of RP which one could designate RPn. Total RP_n is not strictly the sum of the non-random overlap, but because of the small values of these statistics in actual cases, the non-random fraction is the total less the random fraction.

The central concept behind the random-nonrandom distinction can be expressed in a few words: the numbers and geographic distribution of surnames in populations trace the history of the people with the surnames since the origin of the surnames; the distribution of surnames in couples (pairs of parents) indicates the process of structure formation in the temporal cross-section to which the individuals belong. In population studies, those based on a census yield measures subject to low sampling error because the numbers of comparisons of one individual's surname with that of another is large; studies of the surnames of married couples, however, permit large stochastic variation because the number of pairs of surnames of couples is small.

If one has a sample of size N of surnames in members of a breeding population, there are N(N-1) possible comparisons of pairs of surnames at random, but if one compares only the maiden names of the married women with those of their spouses, there are only N/2 comparisons possible. The result of this is much higher degrees of statistical confidence in statements about accumulated structure than in those about process. For instance, in their classic article Crow and Mange (1965) reported that despite a reasonably large value of N the non-random component was not statistically significantly different from zero. The same is true of many other studies including some of mine. This shortcoming has rarely been mentioned in the publications, however. That is because in most studies of surnames the data set encompasses the whole population and authors consider that results are descriptively significant even if not supported by statistical significance. This ignores the fact that surnames are used to sample lines of descent; as models of genetic patterns, even the surnames of all individuals in a population are inherently sampled from a larger universe of descent lines, estimates about which are hence subject to sampling errors which should be estimated.

The only way to gain more information on the process of structure formation than by isonymy is to analyse marriages of couples with different surnames. Devor (1983) applied transition theory to the matrix of the surnames of women as they passed from maiden names to the surnames of their husbands (and hence to their children). The problem Devor encountered is that there were so many different surnames that most possible combinations of two surnames never occurred and most combinations that did occur, occurred rarely. In the kind of societies under discussion, this is always so. For instance, to cite an example (Lasker et al, 1986), among 2,392 marriages recorded in one year in Reading and its vicinity, all but 19 (over 99%) had a surname combination that occurred only once. The way Devor attempted to extract information (and a similar method was independently applied by Pinto-Cisternos et al, 1985) was to combine different surnames. Both these studies did it by merging different surnames which appeared with similar frequency. Tables 1 and 2 give a hypothetical example to show the principle. In the example the first four

Table 1:   Surnames in 16 hypothetical marriages.

Wives'	Husbands'								
	A	B	C	D	E	F	G	H	TOTAL
A. Ayers	0	0	0	0	0	0	0	0	0
B. Begg	0	0	0	0	0	1	0	0	1
C. Cape	0	0	1	0	0	0	0	0	1
D. Dore	1	0	1	0	0	0	0	0	2
E. Evans	0	0	0	0	2	0	1	0	3
F. Fox	0	1	0	0	0	0	0	0	1
G. Green	0	0	0	0	0	2	1	0	3
H. Harris	0	0	0	0	0	0	3	2	5
TOTAL	1	1	2	0	2	3	5	2	16

Table 2:   Surnames in 16 hypothetical marriages:
rare surnames ABCD combined (Ayers, Begg, Cape and Dore);
common surnames EFGH combined (Evans, Fox, Green and
Harris).

Wives'	Husbands'		
	ABCD	EFGH	TOTAL
ABCD	3	1	4
EFGH	1	11	12
TOTAL	4	12	16

surnames (Ayers, Begg, Cape and Dore) are rare and occur one to three times each whereas the latter four surnames (Evans, Fox, Green and Harris) are common and occur four to eight times each. The matrix after merging rare names with rare and common names with common - and after substituting the initial letter for each name - is shown in Table 2. Chi square with one degree of freedom and Yates' correction is 4 and thus statistically significant. The meaning is difficult to adduce, however, since the cell with three entries contains three different combinations of surnames (Cape-Cape, Dore-Ayers and Dore-Cape) and the cell with eleven entries contains six different combinations (Evans-Evans, Evans-Green, Green-Fox, Green-Green, Harris-Green and Harris-Harris).

That is, the matrix methods that have been applied by Devor (1983) and Pinto-Cisternos et al (1985) show that the distribution of pairs of surnames in marriages are non-random and can be interpreted to mean that persons with names common in a community (hence mostly of well-established families) tend to marry others with common names more often than would be expected at random. Likewise, those with rare or unique names are often members of newly in-migrant families and they tend to intermarry with other in-migrant individuals who also have rare names.

These matrix methods are rather crude, however, since they inevitably combine surnames that have different histories into the same category. Thus the surname of an old family line that is dying out, or one just holding its own, may be just as rare as that of a new in-migrant.

REPEATED PAIRS OF SURNAMES (RP)

The alternative strategy which we have proposed is to look at the combinations of specific surnames separately, but then to sum the findings over the whole matrix or over parts of it separately.

In the hypothetical matrix (Table 1) there is one cell with three entries (Harrises married to Greens), three cells with two entries each (Evans-Evans, Green-Fox and Harris-Harris), seven cells with one entry each, and 53 empty cells. If one had more cells with more than one entry or the same number but with larger entries, the number with a single entry would decrease, of course, and the number of empty cells would increase. Although the expected number in almost all individual cells is very small and a chi-square calculation with so many degrees of freedom relative to the number of cases is meaningless, the summary statistics (in this hypothetical example, 53 empty cells versus 11 with entries) can be tested for significance and shown to differ significantly from random. What makes the observed number of occupied cells differ from the expected one is the number of cells with multiple entries: repeated pairs of surnames, we call them. What we have proposed is to count the number of repetitions. If each possible pair of surnames is designated S_{ij}, then the number of repeats may be designated $S_{ij} (S_{ij} - 1)$. That is, each married couple with the ith and j^{th} surnames will be matched with another such couple, $S_{ij} (S_{ij} - 1)$ times. Summed over all pairs of surnames and divided by the number of all possible comparisons:

$$RP = \sum [S_{ij} (S_{ij} - 1)] / [N(N - 1)]$$

in which $N = \sum S_{ij}$.

This measure obviously varies with population size. I am not sure whether it is independent of sample size; but if not it may be necessary to introduce some function of N when samples of different size are compared with each other; this can be done on the basis of the shape of surname frequency distributions as worked out, for instance, by Zei et al (1983), Fox and Lasker (1983), Yasuda (1983), and Yasuda and Saitou (1984). RP for marriages of members of a single lineage is shown to be the same as the coefficient of kinship (k) of persons marrying into the lineage (Lasker, 1988). The weighted mean of one-surname studies of RP equals 4k. Calculated in that way, RP is therefore independent of sample size. Values of one-surname RP can be directly compared with each other and with coefficients of kinship, relationship and inbreeding. However, the samples studied were of similar size and the population structures studied are very different in respect to RP. Furthermore, the ratio of RP to RP_r is independent of sample size because RP_r is calculated on the basis of the identical numbers of each surname as the corresponding RP. In calculation of the random value by repeated runs of a computer model as we have done (Lasker & Kaplan, 1985) and in Chakraborty's (1985) analytic method of calculating RP_r, random selection of mates is accomplished without replacement as if the actual mates were the only potential mates. The alternative of selection with replacement might be more realistic for some studies, but with the sample sizes needed in these studies the method of ascertainment should make very little difference.

By subdividing study samples into subsamples according to exogamy-endogamy, ethnic identity, social status, or other conditions, one can see whether the differences between RP and RP_r in the combined series can be accounted for by any of these social factors.

The hypothetical matrix (Table 1) also demonstrates that it is possible to separate RP into a portion due to isonymy (on the diagonal of the matrix) and a non-isonymous (off-diagonal) portion.   That is, the patterning evident in combinations of surnames in marriage partners is partly ascribable to the kind of inbreeding marked by isonymy (patrilateral mating).   It is identical with clan or lineage endogamy.   The off-diagonal part of the RP may be thought of as interlineage mating systems.   Pairs where two husbands share the same surname and their two wives share the same surname are the only kind of repeated pairs counted in the first study by Lasker and Kaplan (1985).   Off-diagonal cell repeated pairs can involve exchange-like marriage transactions if one uses a programme that equates pairs where a husband and wife have the same surnames as a wife and husband respectively.   If second marriages as well as first ones are included, even the original programme picks up cases of sororate and levirate where a person remarries the sibling of a deceased spouse.   Ethnic subdivisions of a population (such as the British one in which Scots tend to marry Scots and Welsh to marry Welsh) also may lead to elevations of RP.

To some extent these different elements of population structure can be differentiated by dividing the surname matrix into categories and noting which categories of cells contribute more than expected at random to the value of RP. These methods of surname analysis are particularly appropriate for large record-keeping societies, such as Western ones, with long histories of hereditary surnames.   In such societies there is usually an open mode of mate selection in which kinship plays a generally negative role.   The kin-like patterns measured by the extent to which RP exceeds RP_r are thus the result of considerations other than kinship itself.   The extent of these factors can be partialled out by subgrouping the matrix.

We have so far analysed several sets of data by RP.   The first study (Lasker & Kaplan, 1985) is of data from a 100% census of the town of Paracho, Michoacan, Mexico, conducted in 1952.   The main purpose of the census was to study the cottage industries of the community (Kaplan, 1960).   Complete demographic data were also collected.   In Mexico, people are given two surnames, those of both their father and their mother's father, and married women are called by these names rather than by the surname of their husbands. Thus married couples yielded four pairs of surnames (husband's father's by wife's father's, husband's father's by wife's mother's, husband's mother's by wife's father's, and husband's mother's by wife's mother's).   As explained by Shaw (1960) and exploited in subsequent studies of Spanish-speaking peoples (Lasker, 1968; Lasker et al, 1983; Fuster, 1986) this essentially quadruples the quantity of data and permits distinguishing inbreeding due to classificatory cross-cousin mating of two types from parallel matrilateral and parallel patrilateral types. In the case of Paracho, however, we have not placed any emphasis on estimating inbreeding from isonymy because, although the subjects use the Spanish system of naming, many of them had been in the United States and were undoubtedly familiar with the alternative system of surname usage there. If, in speaking with U.S. anthroplogists, even one or two respondents substituted a husband's surname for either of a wife's afternames, the result would be to overstate isonymy and hence exaggerate inbreeding. Such errors would affect RP randomly rather than cumulatively and would be about as likely to reduce RP as to elevate it. They would therefore not introduce a bias. The random noise introduced into RP by an error of this kind would involve only a small fraction of the total number of pairs studied, but the systematic bias of such an error in the case of isonymy would be a large fraction of the isonymous pairs.

In the study of RP in Paracho (Lasker & Kaplan, 1985) 757 pairs of the surnames of husband's father by wife's father were examined, 699 of husband's father by wife's mother, 728 of husband's mother by wife's father, and 681 of husband's mother by wife's mother. For the four categories there were 88, 54, 116 and 102 repetitions respectively, yielding RP ranging from 111×10^{-6} to 220×10^{-6} with a mean of 176×10^{-6}. Random RP was estimated by rearranging the surnames of the wives in random order and then pairing them with the surnames of the males again as if the actual mates were equally probably the mates of any individual of the opposite sex, regardless of differences in age or station in life. Using this standard, the actual number of repeated pairs exceeded the mean randomly expected number by 25%. There is considerable variation among the four types of surname pairs from an excess of 71% above random to a deficiency of 26% but no known pattern of mating preference or source of bias for the variation and we consider it a chance (i.e. unexplained) incidental finding.

However, among these couples over half (365) were Paracho endogamous. For this subset, as one might expect, the frequency of repetitions of pairs of surnames was higher than in the total series. The weighted average for the four types of pairs was 303×10^{-6}. The frequency of repetitions at random in this subset was even higher, however, 332×10^{-6}, so there was no excess of observed over random. The remaining marriages were of three kinds (one person born in Paracho with a spouse from elsewhere, both born in some other place, or each born in a different other place). Numbers of these other types were small, numbers of repeated pairs among them were very small, and expected frequencies were also small.   Calculations of RP for these subsets were therefore subject to large sampling errors and the results are not worth reporting.   On the basis of the Paracho-endogamous subset, however, one can adequately explain the difference between observed and expected RP in the total set as due to the inclusion of the non-Paracho-born individuals.   One should always bear in mind, of course, that the evidence that such an explanation fits the fact does not prove that this was the only possible explanation of the finding. In this instance, however, alternative possible explanations are implausible and the excess of RP over random in the community as a whole seems to be due to the Wahlund effect, a degree of inbreeding through endogamy in a subset of the population.   Since there are four times as many repeated pairs as isonymous pairs, and as isonymy in this population is subject to the possibility of biasing errors, the relationship of observed to random RP, although not directly a measure of inbreeding, seems to give the best estimate of the current (non-random) component of the genetic structure possible from the distribution of surnames among married couples resident there.

    The next attempt to apply the repeated pairs method was on a sample of 2392 marriages in one year in Reading, Wokingham and Henley-on-Thames (Lasker, Mascie-Taylor & Coleman, 1986). Coleman (1980) had found only two cases of isonymy in this sample.   Inbreeding coefficients derived from such small numbers are subject to large sampling errors.   To get the maximum amount of information from RP in this study, the same pair of surnames was considered in reverse order to be repeating pairs, and this modified formulation was called RP2.   In it, for instance, a Smith-Jones couple would be considered a repetition of a Jones-Smith pair.   To calculate RP2 a second list of the surnames was prepared, in which the surnames of the brides were substituted for those of the bridegrooms, and vice-versa, and this list was combined with the correct list.   Repeated pairs occurred twice by this procedure: two Smiths of whatever sex each married to a Jones would appear as Smith-Jones, Smith-Jones, and also as Jones-Smith, Jones-Smith;   also an isonymous pair would appear as, for instance, Brown-Brown, Brown-Brown.   Appropriate ways of counting took this into account.   In the hypothetical matrix (Table 1), if one were to fold it over on the diagonal to show the reciprocal pairs of this type, Begg-Fox would be considered equivalent to Fox-Begg, and these would be added to the number of repeated pairs.

    In the Reading study, RP2 = 3.8×10^{-6}, whereas 12 random runs yield a corresponding value of 2.0×10^{-6}. Thus the observed exceeds random by 88%. A large part of this difference can be explained by the practice of some Sikhs of

Table 3:   Frequency of repetitions of pairs of surnames (RP) in marriages.

	RP x 10^6	random RP x 10^6
Paracho, Mexico (mean)	176	132
Paracho endogamous marriages (mean)	303	332
Reading, England, and vicinity	3.1	1.0*
Sanday, Orkney Islands (mean)	728	623
Sanday residents	914	1024
Sanday, one or both non-resident	427	224
St. Barts Island, French-speaking	7700	7550
Scilly Isles	930	

Calculated by Chakraborty, using his analytic method.

Sources: Lasker and Kaplan (1985); Lasker, Mascie-Taylor and Coleman (1986); Mascie-Taylor, Boyce and Lasker (in preparation); James, Lasker and Morrell (1986); Raspe (unpublished).

using, in lieu of surnames, Singh in the case of males and Kaur in the case of females; There were three such pairs (six repetitions). Furthermore, the influence of ethnic endogamy is further apparent in two marriages in which the same two Chinese surnames occur. One of the other pairs of marriages involving the same two surnames is of a pair of sisters who married each of a pair of brothers at the same place on the same day. In order to make the results comparable to the other studies, Table 3 gives the Reading area figures for RP rather than RP2. In this case the random expectation is given as RP_r by Chakraborty's analytic method.

Other studies applying RP to marital data are planned or under way. Data from Sanday in the Orkney Islands are being analysed by C.G.N. Mascie-Taylor, A. J. Boyce and G. Lasker. Four sets of relationship are being examined: surnames of bridegroom's father and bride's father, bridegroom's father and bride's mother, bridegroom's mother and bride's father, and bridegroom's mother and bride's mother. The data cover the 1210 marriages recorded during the period 1855-1965. Birthplaces are unknown but information on place of residence of the couple make it clear that RP is appreciably reduced if one or both partners is not resident on the island (Table 3). In the case of one parish (Lady) there were 202 marriages in which both parties to a marriage were resident there; among these, RP = 1360 x 10^{-6}, almost the same as RP_r (1399 x 10^{-6}) and higher than for island residents in general.

Another study now under way is of the French-speaking population of a Caribbean island, St. Barts (James et al, 1986). In this inbred population RP is higher than in any other population so far studied (Table 3). Furthermore a high proportion of the repeated pairs of surnames are isonymous - especially isonymous marriages of the commonest surname.

P. Raspe has provided the data for calculation of the overall value of RP among marriages registered in the Scilly Isles in the 19th and 20th centuries (RP $= 93 \times 10^{-5}$). A breakdown into subpopulations and calculation of RP_r is awaited. So are data on Mexican communities being studied by J. M. McCullough.

CONCLUSIONS AND SUMMARY

The data from the different studies now permit one to draw some general hypotheses from the comparison of the different sets of figures:

1.   RP provides more information on marriages than does isonymy, and the more outbred the population, the greater the relative excess of repeated pairs over isonymous ones.

2.   The more traditional a population, the higher the level of RP.

3.   The higher the level of RP, the larger the fraction represented by the random component. That is, as higher levels of RP are encountered, the reason is increasingly explained by accumulated elements of population structure through the high frequency of certain surnames. This does not mean that there are necessarily no absolute increases in the non-random component, however.

4.   The chief factor in high levels of RP is inbreeding - especially selective mating within subgroups (the Wahlund effect).

5.   Factors other than inbreeding are more important, relatively, in less traditional societies.

6.   If mothers' as well as fathers' surnames are studied, there is considerable variation among results from the four types of surname pairs united by marriages.

7.   Second marriages may replicate conditions holding for first marriages. This influence has not yet been explored statistically.

8.   Bride exchange between lineages is less frequent than Kula ring type patterns where mating between lineages tend to repeat in the same direction.

In summary, marital isonymy estimates inbreeding, but most married pairs unite different surname lineages. A method of analysing all combinations of husbands' by wives' surnames is frequency of repeated pairs of surnames (RP).

RP2 adds pairs where names of a husband and his wife match those of a wife and her husband, respectively. These measures can be fractioned into random and non-random components and into segments by geographic, ethnic, social class or name-frequency categories. RP has biological significance but results from social phenomena. Repeated pairs of males of a surname marrying females of a different surname represent a sort of "Kula ring exchange" of mates, with brides passed in one direction among lineages, bridegrooms in another. This produces outbreeding in the ring, inbreeding if there are several distinct rings. In England, RP2 is less than double RP because "bride exchange" between lineages is less common than the "Kula ring" type of pairing; thus this is an open society with pairing somewhat related to outbreeding. Although repeating surname pairs are rare in such societies, the frequency of repetitions is appreciably greater than random expectation and the number of repetitions is much larger than the number of instances of marital isonymy, the traditional measure of mating structure from surnames.

REFERENCES

Chakraborty, R. (1985) A note on the calculation of random RP and its sampling variance. Human Biology, 57: 713-717.

Coleman, D.A. (1980) A note on the frequency of consanguineous marriages in Reading, England, in 1972/73. Human Heredity, 30: 278-285.

Crow, J.F. and Mange, R.P. (1965) Measurement of inbreeding from the frequency of marriages between persons of the same surname. Eugenics Quarterly, 12: (199-203.

Devor, E.J. (1983) Matrix methods for the analysis of isonymous and non-isonymous surname pairs. Human Biology, 55: 257-265.

Ellis, W.S. and Starmer, W.T. (1978) Inbreeding as measured by isonymy, pedigrees and population size in Torbel, Switzerland. American Journal of Human Genetics, 30: 366-376.

Fox, W.R. and Lasker, G.W. (1983) The distribution of surname frequencies. International Statistical Review, 51: 811-87.

Fuster, V. (1986) Relationship by isonymy and migration pattern in Northwest Spain. Human Biology, 58: 391-406.

James, A.V., Lasker, G. and W.T. Morrill (1986) A test of the RP method on St. Bart, F.W.I. (Abstract). Amer.J.Phys.Anthrop. 69: 2(19.

Kaplan, B.A. (1960) Mechanisation in Paracho, a craft community. Alpha Kappa Deltan, 30: 59-65.

Lasker, G.W. (1968) The occurrence of identical (isonymous) surnames in various relationships in pedigrees; a preliminary analysis of the relation of surname combinations to inbreeding. American Journal of Human Genetics, 20: 250-257.

Lasker, G.W. (1985) Surnames and Genetic Structure. Cambridge University Press, Cambridge.

Lasker, G.W. and Kaplan, B.A. (1985) Surnames and genetic structure: repetition of the same pair of names of married couples, a measure of subdivision of the population. Human Biology, 57: 431-439.

Lasker, G.W., Mascie-Taylor, C.G.N. and Coleman, D.A. (1986) The significance for genetic population structure of repeating pairs of surnames in marriages. Human Biology, 58: 421-425.

Lasker, G.W., Wetherington, R.K., Kaplan, B.A. and Kemper, R.V. (1983) Isonymy between two towns in Michoacan. In: Estudios de Antropologia Biologica, pp. 159-163. Universidad Nacional Autonoma de Mexico, Mexico, D.F.

Pinto-Cisternas, J., Castelli, M.C. and Pineda, L. (1985) The use of surnames in the study of population structure. Human Biology, 57: 353-363.

Shaw, R.F. (1960) An index of consanguinity based on the use of the surname in Spanish-speaking countries. Journal of Heredity, 51: 221-230.

Yasuda, N. (1983) Studies of isonymy and inbreeding in Japan. Human Biology, 55: 263-276.

Yasuda, N. and Saitou, N. (1984) Random isonymy and surname distribution in Japan. Biology and Society, 1: 75-84.

Zei, G., Matessi, R.G., Siri, E., Moroni, A. and Cavalli-Sforza, L.L. (1983) Surnames in Sardinia. I. Fit of frequency distributions for neutral alleles and genetic population structure. Annals of Human Genetics, 47: 329-352.

PART III

MEDICAL AND BIOLOGICAL ASPECTS OF INBREEDING

GENETIC RELATEDNESS AND THE
EVOLUTION OF ANIMAL MATING PATTERNS

ANDREW F. READ and PAUL H. HARVEY

*Department of Zoology, University of Oxford,
Oxford, U.K.*

INTRODUCTION

For more than a century, plant and animal breeders have known that matings between very close or very distant relatives can have detrimental effects on offspring fitness (Darwin, 1876; Muller, 1883). Natural selection would be expected to minimise the frequency of such matings if inbreeding or outbreeding depression occurred in the wild. This raises the possibility that there may be some optimal degree of relatedness between potential mates. The idea was mentioned by Sewall Wright in 1933 (see Bateson, 1983), but only recently has its relevance to interpreting animal mating patterns received experimental and theoretical treatment (Bateson, 1978; Shields, 1982).

To demonstrate that inbreeding and outbreeding depression are important in the evolution of mating patterns it must be shown that:

1. inbreeding and outbreeding depression can incur fitness costs in natural populations;

2. individuals avoid matings that would result in these costs;

3. avoidance of close and distant relatives as mates is an adaptation for preventing inbreeding and outbreeding depression, and not just a consequence of some other adaptation.

This chapter summarises the evidence concerning each of these premises. Since this volume is primarily concerned with human mating practices, we shall concentrate on evidence from mammals when possible, and from other vertebrates and even invertebrates or plants when data are not available from higher taxa.

TERMINOLOGY

Inbreeding and outbreeding are matters of degree and do not constitute a dichotomy. Since we are reviewing the possible continuous costs of both inbreeding and outbreeding, it is inappropriate to define a boundary between them.   However, for didactic purposes, we define **close inbreeding** as parent-offspring or full sib matings.   With decreasing relatedness between parents, the costs of inbreeding decrease while the costs of outbreeding increase.

INBREEDING

1.   *The costs*

The offspring of closely related parents often have low fecundity or survivorship.   Such inbreeding depression is generally thought to be a consequence of increased genetic homozygosity which arises because the offspring inherit many of the same genes identical by descent from both parents (Falconer, 1981).   The extent of inbreeding depression resulting from increased homozygosity depends on the genetic structure of the ancestral population.   In outbred populations, new mutations will be selected primarily for their affects in heterozygotes, and recessive mutants will tend to accumulate since they are rarely expressed and therefore rarely selected against.   Most such mutants are detrimental to fitness when expressed as homozygotes.   When mating between close relatives occurs, there will be an increase in homozygosity among the offspring because maternal and paternal genes are likely to be identical by descent from recent common ancestors.   Deleterious recessive genes will then be expressed phenotypically.

There is no doubt that increased homozygosity is associated with inbreeding depression, but it is possible that factors other than the expression of detrimental recessive genes contribute to inbreeding depression.   For example, heterozygote advantage may be involved (Crow & Kimura, 1970; Packer, 1979; Falconer, 1981).   Also, increased homozygosity is associated with increased genetic similarity among siblings, more intense competition for particular resources, and thus lower average fitness (Maynard Smith, 1978).

A vast body of information is available from laboratory, domesticated, agricultural and zoo populations of plants, insects, fish, birds, ungulates, small mammals and primates which confirms the ubiquity and magnitude of inbreeding depression (see Greenwood et al, 1978; Packer, 1979; Ralls & Ballou, 1983 and references therein).   However, much less information is available on inbreeding depression in wild animal populations.   In only two pedigree studies have there been large enough numbers of matings between close relatives, and both are on the same species, but they do confirm that inbreeding depression can occur in wild populations.   Greenwood et al (1978) found that nestling mortality was 27.7% for matings between closely related great tits, but only 16.2% for matings

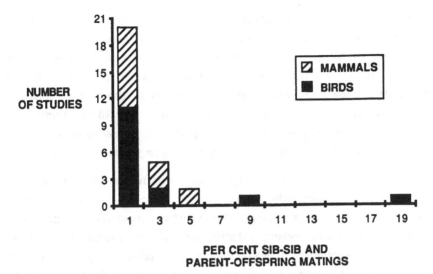

Figure 1: Frequency of close inbreeding found in natural populations of birds and mammals. (From Ralls et al, 1986.)

between less related birds. In a different population of great tits, van Noordwijk and Scharloo (1981) found that the hatching rate of eggs was reduced by 7.5% for every 10% increase in the coefficient of inbreeding. Van Noordwijk and Scharloo (1981) went on to claim that, despite increased nestling mortality, nestling recruitment was not lower for inbred pairs; Greenwood and Harvey (1982) consider that this claim cannot be substantiated given the limited data presented.

2. *Inbreeding avoidance?*

Since there are considerable costs to close inbreeding, strong selection to avoid inbred matings would be expected. Ralls et al (1986) recently reviewed the few pedigree studies of mammals and birds from which reliable rates of close inbreeding can be determined (see Figure 1). In studies of fourteen mammal and thirteen bird populations they found that, in all but two cases, between 0 and 6% of observed matings were sibling–sibling or parent–offspring matings, and that in over half of the studies the figure was less than 2%. The two exceptions were 9.8% in a newly introduced population of European mute swans originating from five adults, and 19.4% in a population of cooperatively breeding splendid wrens.

Despite the low frequency of close inbreeding in natural populations, it does not follow that its incidence is lower than expected by chance. The problem is to define a suitable statistical null hypothesis: what frequency of parent-offspring or full sib matings should we expect if avoidance of close inbreeding does not occur?   This question is difficult to answer directly but, if close breeding *is* avoided, the mechanisms which reduce it should be detectable in natural populations.   Ralls et al (1986) define two categories: (i) *reduction of opportunity* - those mechanisms that reduce the probability of close relatives encountering each other during the mating season;   and (ii) *opportunity not realised* - those mechanisms that reduce the probability of mating between close relatives that do encounter each other during the mating season.

i.   *Reduction of opportunity.*   If close relatives of opposite sex rarely encounter each other during the mating season, the level of close inbreeding will also be rare.   Various processes acting in natural populations result in low encounter rates between related potential mates.   High rates of population turnover due to environmental instability and short life spans may help to reduce the frequency of inbreeding in many small mammal species (Ralls et al, 1986).   For example, in several species of small carnivorous marsupials (*Antechinus* spp.) all males die abruptly at the end of their brief annual breeding season, thereby excluding the possibility of father-daughter incest (Cockburn et al, 1985).

Sex-biased natal dispersal (where males and females usually move different distances from their birth site to breed) also reduces the likelihood that close relatives of opposite sex will encounter each other during the mating season, and is common in both birds and mammals (Greenwood, 1980).   Among gregarious species, such as many primates, one sex tends to remain in the natal group, while the other moves to a different group before breeding.   In less gregarious or solitary species, one sex usually disperses a shorter distance than the other.   Sub-adult males tend to be the predominant dispersers in mammals and sub-adult females in birds (Greenwood, 1980; Greenwood & Harvey, 1982).   In primates, male natal dispersal is the general rule, and breeding individuals in social groups are typically female kin and immigrant males.

Dispersal that occurs between breeding seasons is termed breeding dispersal.   In both birds and mammals, breeding dispersal is almost invariably less frequent and over shorter distances than is natal dispersal.   As with natal dispersal, breeding dispersal tends to be male-biased in mammals and female-biased in birds (Greenwood, 1980; Greenwood & Harvey, 1982).

Hoogland's (1982) study of the black-tailed prairie dog shows just how effective sex-biased dispersal can be for reducing the opportunity for close inbreeding.   Male-biased natal and breeding dispersal removed the possibility of breeding with a close relative for about 90% of females observed in oestrus.

ii.   *Opportunity not realised.*   Even when close relatives encounter each other during the breeding season, inbreeding can still be avoided if relatives recognise and avoid mating with each other. Kin recognition occurs in a variety of birds and mammals (see reviews by Colgan, 1983, and Holmes & Sherman, 1983), and is often based on familiarity during early life. In some species, kin may be recognised even in the absence of prior association. The mechanisms used to recognise kin vary between species.

Detailed long-term studies of the behaviour shown by sexually mature individuals towards kin and non-kin are relatively rare, but most of the available evidence indicates that individuals avoid mating with close relatives and that this avoidance is not just a question of dispersing and then mating at random (c.f. Moore & Ali, 1984, but see van Noordwijk et al, 1985). Hoogland (1982) reports eight cases in which female black-tailed prairie dogs reduced the probability that their offspring would be sired by closely related males.   The females either failed to come into oestrus, failed to copulate while in oestrus, copulated with only unrelated males in their groups, left the group to mate with an unrelated male in another group and then returned, or copulated with their relatives and with unrelated males who temporarily invaded the group.

Chimpanzees are one of the few mammals in which females disperse and males often live in the same group as their mothers. Pusey (1980) recorded that the rates of intra-troop associations between males and immature females (usually maternal siblings) dropped abruptly when the females commenced cycling.   Sexual behaviour between maternal siblings was very infrequent, apparently due to a lack of interest in their relatives by both males and females: the males seldom courted but, when they did, the females rarely responded. And during the rare occasions when copulation occurred, the females usually screamed more often than they did when copulating with an unrelated male. Furthermore, sexually mature sons did not copulate with their mothers, even though many unrelated males did.

Similar evidence of mechanisms, other than dispersal, which result in decreased levels of inbreeding have been documented in many other species of mammals and birds (see Ralls et al, 1986).

In summary, the most widely documented mechanisms found in animal populations which have the effect of reducing the level of close inbreeding are sex-biased dispersal and, when relatives do meet, avoidance of copulation by one or both sexes.

3.   *Does inbreeding depression influence mate choice?*

Unless the mechanisms outlined above can be demonstrated to be adaptations which have been selected for because they avoid inbreeding depression, the possibility remains that such behaviour patterns may have evolved for other reasons.   Mechanisms which inhibit copulations between kin when they meet, such as active avoidance of copulation and failure to come into oestrus, are difficult to explain unless they are maintained by selection to prevent close inbreeding. But the ultimate (evolutionary) causes of dispersal are the centre of some controversy (e.g. Shields, 1982; Moore & Ali, 1984; Packer, 1985; Dobson & Jones, 1985).   It is clear that much dispersal occurs for reasons other than inbreeding avoidance.

Among birds there is no evidence that *breeding* dispersal reduces the level of inbreeding (Ralls et al, 1986).   And there are several benefits which can be gained by dispersal (and hence outbreeding):   dispersing individuals may find better or unexploited resources and at the same time minimise the negative effects of intraspecific competition, particularly for mates (Horn, 1978; Waser & Jones, 1983; Moore & Ali, 1984; Ralls et al, 1986).   For example, those male rhesus monkeys with the highest frequencies of copulation are the ones most likely to transfer between groups during the breeding season.   There is also a tendency in some species for breeding dispersal to be from groups with few available females to groups with many (Packer, 1979; Greenwood, 1980).   Such behaviour patterns appear to increase mating opportunities, and could therefore be selected for even in the absence of inbreeding depression. Furthermore, at least some male emigration is the result of forcible expulsion by other males in the group and the expelled males spend much of their time trying to get back into any bisexual group (Moore & Ali, 1984).   Young male olive baboons, for example, often receive considerable harassment in their natal groups, are forcibly evicted, and remain outside bisexual troops until they can enter a new troop (Packer, 1979).

These alternative hypotheses undoubtedly account for some dispersal in some species, but they fail to account for many instances of dispersal.   We use several examples to illustrate the point.   First, some male primates transfer directly from their natal group to another even though they receive more aggression from members of the new troop than the old (see Packer, 1985). According to the intraspecific competition hypothesis, we should expect males to stay in their natal groups under such circumstances.   Second, there is no apparent female competition for mates in the polygynous pied flycatcher, yet this   species   exhibits   female-biased   dispersal   (Greenwood,   1980);   the intraspecific competition hypothesis predicts male dispersal. Our third example is breeding dispersal by male Belding's ground squirrels which coincides with maturation of their probable daughters, and the most polygynous males (which have a greater chance of having daughters in the adjacent groups) move the farthest (Sherman, 1981). The inbreeding avoidance hypothesis can explain each of the above cases.

There is often voluntary natal dispersal by large, apparently healthy and dominant individuals (Howard, 1960): dominant male vervet monkeys or black-tailed prairie dogs may sometimes abdicate and transfer from a group after several years (Henzi & Lucas, 1980; Hoogland, 1982). Moore and Ali (1984) suggest that such males may be "bet-hedging" (i.e. decreasing the risk of offspring extinction by having offspring in several groups), but at least in the black-tailed prairie dog, males are more likely to move to a new breeding group if they have adult daughters present in their own group (Hoogland, 1982). And, in some communal bird species, inbreeding avoidance is not simply a demographic consequence of dispersal (Greenwood & Harvey, 1982). Young acorn woodpeckers of either sex, for example, disperse from their natal territory even if their opposite-sexed parent is still in residence but lacks a mate (Koenig & Pitelka, 1979).

The alternative hypotheses also fail to fully explain why so many species of birds and mammals exhibit sex-biased natal dispersal. There may be several reasons other than the avoidance of inbreeding depression which help explain the evolution of the sex bias. There are many sex differences that could lead to differential selection pressures on the dispersal of the sexes, such as body size, foraging behaviour and level of parental investment (Ralls et al, 1986). Waser and Jones (1983) suggest that the direction of the sex bias correlates with the extent to which the offspring of each sex are a threat to the reproductive success of each parent. They also discuss the possibility that parents can best increase the reproductive success of the sex that does not disperse.

But perhaps the strongest evidence that sex-biased dispersal functions as an inbreeding avoidance mechanism comes from studies by Cockburn et al (1985) on two species of *Antechinus*, small Australian marsupials. Females remain in their natal area but all the juvenile males disperse, attempt to breed, and then die abruptly at the end of the short annual mating season. Male dispersal occurs even in those litters which consist of all males and in those with only one. Intrasexual competition for mates can be excluded as the cause of dispersal since there is no opportunity for interaction between fathers and sons, and juvenile males disperse even if they are the only males in their litter. Mothers apparently cause the dispersal of their sons and recruit unrelated juvenile males to live with themselves and their daughters. Because this does not generally lead to a reduction in the number of individuals in the nest, competition for resources can be excluded as a selective agent. The only benefit accruing to a mother by exchanging sons for other males with whom her daughters will mate appears to be the avoidance of inbreeding depression among her grandchildren.

A similar but less conclusive example which appears to contradict the intrasexual competition hypothesis is the Californian ground squirrel: all males disperse from their natal group, including those that mature in colonies supplied with supplementary food and those which contain no other resident males (Dobson, 1979).

OUTBREEDING

1.   *The costs*

Outbreeding depression is generally believed to result from the disruption of successful genomes ("coadapted gene complexes" - Dobzhansky, 1948; Templeton et al, 1986), and is manifested as decreased fitness in the F1 or later generations. It is hypothesised that genes at many loci are selected for their joint effects and, if these are disrupted or individual genes required for a particular adaptation are lost during recombination, the resulting offspring will not be adapted to either of the parents' environments (Price & Waser, 1979; Shields, 1982). Furthermore, problems of genetic incompatibility of parents may lead to zygotic and embryonic mortality, still births, decreased fertility and increased juvenile mortality (Falconer, 1981; Shields, 1982; Bateson, 1983; Templeton et al, 1986).

The evidence for outbreeding depression rests almost entirely on the offspring produced by very distantly related parents from geographically distinct populations.

The only examples of outbreeding depression on a more local scale appear to be those shown in three perennial herbs which have relatively short mean dispersal distances: *Delphinium nelsoni* (Price & Waser, 1979), *Ipomopsis aggregata* (Waser & Price, 1983), and *Castilleja miniata* (Lertzman, 1981; see Waser & Price, 1983). The most widely cited of these is Price and Waser's (1979) study of *Delphinium nelsoni*, in which flowers were artificially pollinated by plants from various distances (distance was assumed to be directly correlated with the degree of relatedness of the parent plants), and the effects on fitness were measured. When compared with self-pollination or pollination by plants 1000 m away, pollination by plants between 1 and 100 m, and centred on about

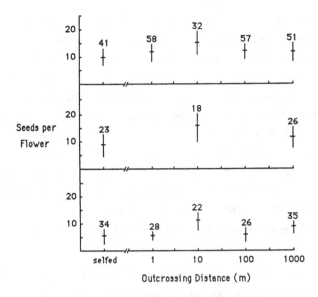

Figure 2: Seed set as a function of geographical distance separating the parents in three trials of artificially pollinated *Delphinium nelsoni*. Values are means ±95% confidence units, with sample sizes (number of plants pollinated) above. 10 m seed sets were significantly higher overall than either 1000 m (P<0.05) or selfed (P<0.005) seed sets (after Price & Waser, 1979).

10 m away, resulted in significantly increased numbers of seeds produced per flower (Figure 2) and significantly increased subsequent seedling viability. Price and Waser controlled for ecological factors such as sibling competition and overcrowding, and they concluded that the decreased fitnesses were due to the genotypic imbalances of the various crosses. The magnitude of inbreeding depression was considerably greater than that of outbreeding depression, despite the fact that inbreeding depression occurred over 10 m while outbreeding depression was over 900 m from the parent plant. In short, Price and Waser (1979) did demonstrate outbreeding depression, but the genetic costs of outbreeding seemed to increase very slowly with decreasing relatedness. Very similar patterns were found in the two other plant studies cited above.

There is no equivalent evidence for outbreeding depression on a local scale in animals. However, some studies do claim to have identified evidence for the existence of coadapted gene complexes. These are either laboratory studies or hybrid crosses between subspecies and/or geographically distinct races. Examples are reviewed by Endler (1977), Shields (1982) and Templeton et al (1986). We shall discuss two of these here to show the sort of evidence involved and the problems associated with generalising from it.

Drosophila mercatorum is normally an outcrossing sexually reproducing species, but parthenogenesis can be induced in virgin females, thereby developing sexually reproducing homozygous strains after a single generation. Some strains survive and others become extinct; those that survive are presumably the ones with well-adapted homozygous genotypes. When different strains are subsequently hybridised, there is a massive decrease in offspring viability correlated with the degree of hybridisation (Templeton et al, 1986). From this we can conclude that two successful homozygous genomes have lowered fitness when combined in heterozygotes. Co-adaptation in the sense of epistatic gene interactions are not necessarily involved, e.g. some alleles may need to be present as homozygotes in order to produce a gene product; those existing in homozygous strains would thus not have any phenotypic effect in the hybrid crosses.

Our second example is of three small terrestrial Australian frog species of the genus *Pseudophryne* which are said to show outbreeding depression where they hybridise (Templeton et al, 1986). The three species are largely allopatric, but there are narrow hybrid zones where their ranges overlap. In these zones embryonic mortality is higher and is correlated with hybridisation between strains (Woodruff, 1979).

Examples of hybridisation between races and species cited as evidence of outbreeding depression (Shields, 1982; Templeton et al, 1986), such as those outlined above, may indeed result in decreased offspring fitness. But we disagree with Shields (1982) that "there is no compelling reason, logical or empirical, to believe that the same processes and resulting incompatibilities (shown by such hybridisations) could not occur in more closely related, but still differentiated forms (i.e. between families in a group)" (p. 91). Clearly such differentiation requires genetic isolation of such "family" groups within a population's gene pool (Waser & Price, 1983). Until such a restricted gene flow has been demonstrated on a local scale in animal populations we can see no theoretical or empirical reason to extrapolate the effects of intraspecific hybridisation of individuals from geographically distant populations or subspecies (cf. Shields, 1982, p. 91) to intrapopulation matings.

The evidence for outbreeding depression in the wild is thus very poor, unless interracial or interspecific hybrids are involved and these may be of little relevance to understanding the evolution of local mating patterns, except, perhaps, at the boundaries between subspecies or incipient species.

2.   *Outbreeding avoidance?*

If, as Shields (1982) and others have argued, avoidance of outbreeding depression is an important selective pressure on mating patterns, then we would expect animals to avoid outbreeding. In theory, outbreeding could be avoided by not dispersing any further than necessary, and by recognising distantly related animals and then avoiding copulations with them.

It is very difficult to show whether animal dispersal patterns have evolved to minimise outbreeding. It is known that the frequency distribution of animal dispersal distances from their birth sites is usually highly skewed, with most animals moving relatively short distances and a few moving long distances (reviewed by Shields, 1982). However, in the absence of any null hypothesis about what constitutes random dispersal, the skewed frequency distribution cannot be taken as evidence that animal dispersal patterns have evolved to minimise outbreeding (cf. Shields, 1982).

The evidence for active avoidance of distantly related animals in the wild is very meagre, and is often in fact negative. Female olive baboons, for example, often seem to prefer strange males (Packer, 1979). And none of the three Australian frog species discussed above show any avoidance of interspecific matings, despite the associated fitness cost (Woodruff, 1979). However, laboratory studies of three species show that, given the choice, animals may prefer to associate with individuals that are not too distantly related: zebra finches (Miller, 1979; Slater & Clements, 1981), house mice (D'Udine & Alleva, 1983) and Japanese quail (Bateson, 1978). However, there are problems with these studies, such as not measuring actual *mating* preferences (Japanese quail - Bateson, 1978; zebra finches - Miller, 1979 but see Slater & Clements, 1981) or offering potential mates with only a restricted range of relatednesses (zebra finches, Slater & Clements, 1981). But they do suggest that animals of some species may recognise and then avoid distantly related animals.

We do not know of any field evidence that supports the idea that birds and mammals prefer to mate with fairly close relatives (say cousins) over more distantly related animals.   Van Noordwijk et al (1985) could find no discrimination on the basis of relatedness in the process of pair formation in a population of great tits.

There is one example in which evolution appears to have favoured mechanisms which decrease the frequency of matings between distantly related animals, so as to avoid breaking up locally adapted gene complexes.   Several species of moth living in deciduous North American forests time the hatching of eggs so that the larvae appear at the same time as the leaves, which maximises larval survivorship.   However, individual trees leaf at different times and, it appears, different individual moths within a species have become adapted to the timing of foliation of the trees on which they live.   This local adaptation is maintained by a large proportion of flightless individuals and subsequently very low rates of dispersal, together with asexual reproduction (Mitter et al, 1979; Schneider, 1980).   Sexual forms of the same moth species do not show the same degree of local adaptation.

3.   *Does outbreeding depression influence mate choice?*

Even if outbreeding avoidance was demonstrated in natural populations of animals, other costs associated with outbreeding may well be more important evolutionary forces influencing mate choice than outbreeding depression itself. For example, dispersal may be costly (see Gains & McClenaghan, 1980, and Waser & Jones, 1983).   Lone animals lose the antipredator and foraging advantages of group living.   Furthermore, dispersing animals often traverse inhospitable habitats, with increased risks of starvation and predation, with the latter possibly increased through lack of familiarity with escape routes. Greater mortality in dispersers compared with non-dispersers has been implicated in birds (e.g. blackbirds), non-primate mammals (e.g. water voles) and primates (e.g. olive baboons).   Dispersing individuals are often in relatively poor condition; dispersing rabbits and black bears, for example, have a slower growth rate than do non-dispersing individuals.   Abundant evidence suggests that dispersers may face considerable difficulties when joining an established population, because residents (particularly same-sex residents) often respond aggressively to strangers.   And when immigrants are allowed to settle they may enter at the bottom of the social hierarchy.

Other hypothetical costs of dispersal which are less well documented in wild populations, include greater risk of infection by novel pathogens (Bateson, 1983; Moore & Ali, 1984), increased costs of mate choice, particularly if mate density is low (Partridge, 1983), loss of reproductive time (Bengtsson, 1978), decreased foraging efficiency in unfamiliar habitats (Moore & Ali, 1984), and any investment which may be required in a new territory, such as burrow construction (Waser & Jones, 1983).   Because of these costs of dispersing, natural selection is likely to favour individuals that minimise the distance they disperse, even in the absence of outbreeding depression.

There are other costs associated with outbreeding which are not related to outbreeding depression, but which may lead to outbreeding avoidance. Various authors have suggested that animals may decrease their inclusive fitness by outbreeding, for reasons other than outbreeding depression, and should therefore be selected to inbreed (Alexander, 1974; Bengtsson, 1978; Smith, 1979). In its simplest form, the argument runs that individuals breeding with close relatives can increase inclusive fitness if one sex competes for mates. For example, if males compete for females but males do not invest in the resulting offspring, then a female can increase the mating success of a male relative, and hence her own inclusive fitness, if she mates with him (Maynard Smith, 1978; Smith, 1979; Packer, 1979). Polygynous mammals, it is hypothesised, may have greater frequencies of inbred matings (e.g. fallow deer: Smith, 1979), although Ralls and Ballou (1983) list several species in which this does not seem to occur. Increased inclusive fitness may explain sibling matings in a species of hunting wasp (Cowan, 1979). For such arguments to hold true, decreased fitness of the inbred offspring must be compensated for by the increase in inclusive fitness to both parents.

Inclusive fitness may also be increased by having close kin as neighbours, since cooperation and altruism are most likely to develop between relatives than non-relatives (Wilson, 1976; Greenwood et al, 1979). In several squirrel species, muntjacs and feral domestic cats, decreased levels of aggression have been recorded between closely related neighbouring females compared to those between more distantly related females (Waser & Jones, 1983). Defence of common areas and more frequent predator warnings by cooperating kin have also been recorded.

Therefore, if outbreeding avoidance could be detected in natural populations, it need not have evolved as a result of selective pressures minimising the effects of outbreeding depression. And, even if outbreeding depression itself is demonstrated in natural animal populations, the costs of dispersal and the benefits to inclusive fitness of mating with more closely related animals could well be more important evolutionary forces shaping mating patterns.

DISCUSSION

For no one species do we have a clear idea of the costs associated with matings between individuals of varying degrees of relatedness, yet these are likely to vary widely between populations. The relative costs involved will depend on, among other things, the previous genetic structure of the population (Bateson, 1983; Templeton, 1986). Once outbreeding has been adopted deleterious recessive alleles will accumulate and inbreeding depression will be greater.   Similarly, if inbreeding is adopted, then deleterious recessive alleles will be lost, and the costs of inbreeding decreased.   There is also likely to be large interpopulation variation in the potential inclusive fitness gains of inbreeding and in the costs of dispersal.

Given this variability, and our lack of knowledge of how the various costs interact, are inbreeding and outbreeding depression factors involved in the evolution of animal mating patterns?   Inbreeding depression is almost certainly involved, at least in some species.   The genetic costs of inbreeding are well documented, many animals do avoid close inbreeding, and in several cases this avoidance can be explained only as an adaptation for avoiding inbreeding depression itself.   There is, however, no evidence from natural populations of animals that outbreeding depression is an important factor, except in matings between species, subspecies and widely divergent geographic races.

What then of the studies which show fine tuned 'mating' preferences in Japanese quail, zebra finches and mice discussed above?   In the absence of conclusive evidence about the existence of outbreeding depression on a local scale, it seems premature to invoke the avoidance of outbreeding depression as an explanation.   Alternative hypotheses should also be considered.   For example, the preferred degree of relatedness may reflect a trade off which minimises inbreeding depression without negating the potential inclusive fitness benefits of inbreeding.

CONCLUSION

Inbreeding depression is a factor in the evolution of mating patterns, at least in some species.   There is, however, very little evidence to show that the avoidance of outbreeding depression is involved except, perhaps, at the level of matings between different geographical races, subspecies or species.

ACKNOWLEDGMENTS

Andrew Read was supported by an A.C.U./British Council Commonwealth Scholarship during the period of this study.

REFERENCES

Alexander, R.D. (1974).   The evolution of social behaviour.   Annual Review of
    Ecology and Systematics, **5**, 325-383.
Bateson, P. (1978).   Sexual imprinting and optimal outbreeding.   Nature, **273**,
    659-660.

Bateson, P. (1983). Optimal outbreeding, In: P. Bateson (ed.), Mate Choice, pp. 257-278. Cambridge: Cambridge University Press.

Bengtsson, B.O. (1978). Avoid inbreeding: at what cost? Journal of Theoretical Biology, **73**, 439-444.

Cockburn, A., Scott, M.P. & Scotts, D. (1985). Inbreeding avoidance and male biased breeding dispersal in *Antechinus* sp. Animal Behaviour, **33**, 908-915.

Colgan, P. (1983). Comparative Social Recognition. New York: Wiley.

Cowan, D.P. (1979). Sibling mating in a hunting wasp: adaptive inbreeding. Science, **205**, 1403-1405.

Crow, J.F. & Kimura, M. (1970). An Introduction to Population Genetics Theory. New York: Harper & Row.

Darwin, C. (1876). The Effects of Cross and Self Fertilization in the Vegetable Kingdom. London: Murray.

Dobson, F.S. (1979). An experimental study of dispersal in the California ground squirrel. Ecology, **60**, 1103-1109.

Dobson, F.S. & Jones, W.T. (1985). Multiple causes of dispersal. American Naturalist, **126**, 855-858.

Dobzhansky, T. (1948). Genetics of natural populations. XVIII Experiments on chromosomes of *Drosophila pseudoobscura* from different geographical regions. Genetics, **33**, 588-602.

D'Udine, B. & Alleva, E. (1983). Early experience and sexual preferences in rodents, In: P. Bateson (ed.), Mate Choice, pp. 311-330. Cambridge: Cambridge University Press.

Endler, J.A. (1977). Geographic Variation, Speciation and Clines. New Jersey: Princeton University Press.

Falconer, D.S. (1981). An Introduction to Quantitative Genetics, 2nd edn. London: Longman.

Gains, M.S. & McClenaghan, L.R. (1980). Dispersal in small mammals. Annual Review of Ecolgy Systematics, **11**, 163-196.

Greenwood, P.J. (1980). Mating systems, philopatry and dispersal in birds and mammals. Animal Behaviour, **28**, 1140-1162.

Greenwood, P.J. & Harvey, P.H. (1982). The natal and breeding dispersal of birds. Annual Review of Ecology and Systematics, **13**, 1-21.

Greenwood, P.J., Harvey, P.H. & Perrins, C.M. (1978). Inbreeding and dispersal in the great tit. Nature, **271**, 52-54.

Greenwood, P.J., Harvey, P.H. & Perrins, C.M. (1979). Kin selection and territoriality in birds? A test. Animal Behaviour, **27**, 645-651.

Henzi, S.D. & Lucas, J.W. (1980). Observations of the inter-troop movement of adult vervet monkeys (*Cercopithecus aethiops*). Folia Primatologica, **33**, 220-235.

Holmes, W.G., Sherman, P.W. (1983). Kin recognition in animals. American Scientist, **71**, 46-55.

Hoogland, J.L. (1982). Prairie dogs avoid extreme inbreeding. Science, **215**, 1639-1641.

Horn, H.S. (1978). Optimal tactics of reproduction and life history, In: J.R. Krebs & N.B. Davies (eds.), Behavioural Ecology: An Evolutionary Approach, pp. 411-429. Oxford: Blackwell.

Howard, W.E. (1960). Innate and environmental dispersal of individual vertebrates. American Midland Naturalist, **63**, 152-161.

Koenig, W.D. & Pitelka, F.A. (1979).   Relatedness and inbreeding avoidance: counterploys in the communally nesting acorn woodpecker. Science, **206**, 1103-1105.

Lertzman, K.P. (1981).   Pollen transfer: processes and consequences. M.Sc. thesis, University of British Columbia, Vancouver.

Maynard Smith, J. (1978).   The Evolution of Sex.   Cambridge: Cambridge University Press.

Miller, D.B. (1979).   Long term recognition of father's song by female zebra finches. Nature, **280**, 389-391.

Mitter, C., Futuyma, D.J., Schneider, J.C. & Haire, J.D. (1979).   Genetic variation and host plant relations in a parthenogenic moth. Evolution, **33**, 777-790.

Moore, J. & Ali, R. (1984).   Are dispersal and inbreeding avoidance related? Animal Behaviour, **32**, 94-112.

Muller, M. (1883). The Fertilisation of Flowers. London: MacMillan.

Noordwijk, A.J.van & Scharloo, W. (1981).   Inbreeding in an island population of the great tit. Evolution, **35**, 674-688.

Noordwijk, A.J.van, Tienderen, P.H.van & Jong, G.de (1985).   Genealogical evidence for random mating in a natural population of the great tit (*Parus major* L.). Naturwissenshaften, **72**, 104-105.

Packer, C. (1979).   Inter-troop transfer and inbreeding avoidance in *Papio anubis*. Animal Behaviour, **27**, 1-36.

Packer, C. (1985). Dispersal and inbreeding avoidance. Animal Behaviour, **33**, 676-678.

Partridge, L. (1983).   Non-random mating and offspring fitness, In: P. Bateson (ed.), Mate Choice, pp. 227-256.   Cambridge: Cambridge University Press.

Price, M.V. & Waser, N.M. (1979).   Pollen dispersal and optimal out-crossing in *Delphinium nelsoni*. Nature, **277**, 294-297.

Pusey, A.E. (1980).   Inbreeding avoidance in chimpanzees.   Animal Behaviour, **28**, 543-552.

Ralls, K. & Ballou, J. (1983).   Extinction: lessons from zoos. In, C. M. Schonewall-Cox, S. M. Chambers, B. MacBryde & L. Thomas (eds.), Conservation Genetics: A Reference for Managing Wild Animal and Plant Populations, pp. 164-184. London: Benjamin/Cummings.

Ralls, K., Harvey, P.H. & Lyles, A.M. (1986).   Inbreeding in natural populations of birds and mammals,   In: M. Soule (ed.),   Conservation Biology: Science of Diversity. Michigan University Press (in press).

Schneider, J.C. (1980).   The role of parthenogenesis and female aptery in microgeographic, ecological adaptation in the fall cankerworm, *Alsophila pometaria* Harris (Lepidoptera: Geometridae). Ecology, **61**, 1082-1090.

Sherman, P.W. (1981).   Kinship, demography and Belding's ground squirrel neoptism. Behavioural Ecology and Sociobiology, **8**, 251-259.

Shields, W.M. (1982).   Philopatry, Inbreeding and the Evolution of Sex. Albany: State University of New York Press.

Slater, P.F.J. & Clements, F.A. (1981).   Incestuous mating in zebra finches. Zeitschrift fur Tierpsychologie, **57**, 201-208.

Smith, R.H. (1979).   On selection for inbreeding in polygynous animals. Heredity, **43**, 205-211.

Templeton, A.R. (1986). Inferences on natural population structure from genetic studies on captive mammalian populations. MS.

Templeton, A.R., Hemmer, H., Mace, G., Seal, U.S., Shields, W.M. & Woodruff, D.S. (1986). Local adaptation, coadaptation and population boundaries, In: K. Ralls & J. Ballou (eds.), Proceedings of the Workshop on Genetic Management of Captive Populations. Zoo Biology, 5(2).

Waser, P.M. & Jones, W.T. (1983). Natal philopatry among solitary mammals. Quarterly Review of Biology, **58**, 355-390.

Waser, P.M. & Price, M.V. (1983). Optimal and actual outcrossing in plants, and the nature of the plant-pollinator interaction, In: C. E. Jones & R. J. Little. New York: Van Nostrand Reinhold.

Wilson, E.O. (1976). The central problems of sociobiology, In: R. M. May (ed.), Theoretical Ecology: Principles and Applications, pp. 205-217. Oxford: Blackwell.

Woodruff, D.S. (1979). Post mating reproductive isolation in Pseudophryne and the evolutionary significance of hyprid zones. Science, **203**, 561-563.

Wright, S. (1933). The roles of mutation, inbreeding, crossbreeding and selection in evolution, In: D. F. Jones (ed.), Proceedings of the VIth International Congress on Genetics, vol. 1, pp. 356-366. New York: Brooklyn Botanical Garden.

REGULATION OF
MATING CHOICE IN NONHUMAN PRIMATES

D. QUIATT

*Department of Anthropology, University of Colorado at Denver,
Denver, U.S.A.*

INTRODUCTION: CHANGING VIEWS OF NON-HUMAN
PRIMATE SOCIAL ORGANISATION

The notion prevailed until not long ago that primates of most species lived in promiscuous, inbred groups in which a dominance hierarchy of competing, reproductively active males provided the "main axis" of social organisation. This was in striking contrast with our own species, in which genealogical systems, marriage rules, and the incest taboo regulate outbreeding in ways which promote cooperation both within and between groups. In that landmark collection of primatological field reports edited by DeVore (1965), Carpenter's contribution contains the only explicit discussion of gene dispersal via emigration, and even Carpenter evidently assumed that emigration and immigration were characteristic of groups in social "disequilibrium" and, in that sense, not regular features of individual life history (Carpenter, 1965). Intergroup transfers by Old World monkeys are either not reported at all - Jay (1965) observes somewhat ambiguously that langurs ".... rarely change groups. The membership of a group remains constant except for births, deaths, and the departure of a few adult males that leave to live as nongroup males....." - or are presented as cases both exceptional and highly consequential for social order, as indeed they sometimes may be: Hall and DeVore (1965) thus relate in detail the consequences of two transfers by male baboons, one of them observed by Washburn. Most surprising, in retrospect, is Carl Koford's failure to record intergroup transfers among the rhesus monkeys of Cayo Santiago, Puerto Rico. By 1963, when contributors to this collection met in conference, Koford must have suspected that the normal course for all males was to initiate and pursue reproductive life elsewhere than in the natal group; yet he noted only that ".... most adolescent males leave their mothers and often their natal group to become peripheral males" (Koford, 1965).

In these "early modern" field studies, description of species-characteristic behaviour and comparison across species were first objectives; for such purposes the day-to-day foraging group appeared to constitute a natural unit of study, varying in ways which suggested evolutionary adaptation to different habitats.

Summary descriptions also served to document general assertions about within-group functions of social behavior, and there was a tendency to frame them in terms of typical relations between age-sex classes (almost always adult male/adult female or mother/offspring), with little attention to individual variation within class or to behaviour in extra- and intergroup contexts. For anthropologists, this tendency may reflect the influence of the social anthropologist A. R. Radcliffe-Brown, whose theoretical interests bore on the cohesive structure of the social group rather than on the varying interests of individuals and who, it has been suggested, ".... influenced both data collection methods and the general orientation of the first physical anthropologists to do field studies on primate social behavior" (Richard, 1985a, citing Gilmore, 1981).

Once information had been accumulated on different groups of the same species and on the same groups over a time-span of a generation or more, research interest gradually shifted to behaviour in process (i.e. to aspects of sequence, duration, intensity, etc.), to relations between subgroups, and to specific actions of individuals of known genealogy. In studies of social behaviour, the individual animal is necessarily the unit of observation and initial description; today the individual is likely to be the primary unit of analysis as well, with attention to variability within and between classes of individuals.

Nevertheless, summary descriptions of non-human primate social organisation still take the spatially cohesive foraging group (= troop, band, herd) as their starting point. Primatologists usually distinguish 5 or 6 categories of "social organisation", as in Table 1. The utility of recognising inclusive classes of social organisation seems confirmed by agreement, while at the same time agreement is probably furthered by recognition of the utility of any rough and ready classification for purposes of instruction and general communication. Thus a tendency to reify the group at a particular level of spatial inclusiveness is perpetuated, along with buried assumptions as to internal unity and homogeneity of processes.

ANIMAL MATING SYSTEMS THEORY

Common assumptions underlay early correlational schemes of primate social organisation (see Richard, 1985a, for a comprehensive review and critique) and the developing broader theory of animal mating systems in evolution - i.e. that social behaviour is influenced by habitat and that the comparative approach is essential to understanding behaviour as adaptation. However, the goal of current mating system theory, firmly grounded in

Table 1: Three classifications of non-human primate social organisation

Jolly & Plog, 1979	Chalmers, 1979	Gouzoules, 1984	Typical mating patterns
Noyau	Solitary species	Solitary	Polygyny/promiscuity
Territorial pair	Monogamous family groups	Family groups	Monogamy
One-male group	One-male (uni-male) groups	One-male units (harems)	Polygyny
Multi-male troop (includes chimp-anzees as a sub-category)	Multi-male groups (includes age-graded groups, e.g. gorilla)	One-male units within multi-male groups	Polygyny
One-male groups within multi-male troop	Difficult to classify *Papio hamadryas* *Theropithecus gelada* *Pan troglodytes* *Pongo pygmaeus*	Multi-male, multi-female	Promiscuity

behavioural ecology, is to integrate explanations of mating system types at the stage of initial hypothesis formation through consistent focus on individual-individual competition for resources and on differential sex-based strategies for competition, not, as in the older approach, via post hoc interpretation of correlations between habitat and social organisation.

The notion of individual cost-benefit decision-making is central to mating system theory, the argument being that although an animal's decisions may or may not be based on rational consideration of costs, benefits, and uncertainties associated with alternative behavioural options "... they can be examined as if they were" (Wittenberger, 1979). The common currency of these decisions is lifetime reproductive output; overlapping interests of genetic relatives and a partial conflict of interest between males and females provide the fundamental dynamics. The problems with thinking about behaviour in economic terms are compounded when "economic man" is translated into an evolutionary setting, but there is no doubt that such thinking provides a useful tool for "... comprehending the often complex behavioral strategies involved in animal mating systems" (Wittenberger, 1979).

The complexity of mating strategies is evident in Wittenberger's discussions of mating types. Building on terminology developed by Lack (1968) and Selander (1972), he breaks the general types *monogamy, polygyny, polyandry* and *promiscuity* into subcategories according to spatial and temporal characteristics.   *Monogamy* yields three spatial subcategories: *Territorial* (the pair defends a common territory), *Female-Defence* (each male defends access to a female), and *Dominance-Based* (females maintain pair bonds by dominating subordinate females), and two temporal subcategories: *Serial* (a new mate each year or each breeding cycle) and *Permanent* (usually for life, following successful breeding).   *Polygyny, Polyandry* and *Promiscuity* are similarly subdivided (*Promiscuity* by definition has no patterned continuity of reproductive relationships, hence no temporal subcategories) for a total of 18 alternative behaviour forms (Wittenberger, 1979).

Differences in male and female strategies of competition for resources suggest a general logic of gene dispersal in mammalian species, in which young characteristically are most dependent on female parents.   Female competition is likely to centre on resources crucial to offspring care, and natural selection should favour females who distribute themselves, individually or in groups with overlapping reproductive interests, for maximum efficiency in exploiting those resources;   male competition is more likely to centre on competition for mates, and distribution of males therefore should be regulated by distribution of females (Greenwood, 1980;   Wittenberger, 1980;   Richard, 1985a).

In most species of primates the typical mammalian pattern prevails: males leave the natal troop at reproductive maturity, while females remain in company with close, predominantly female relatives.   Wrangham has argued that, if females compete primarily for food resources and males compete primarily for mates, where foods are found in large, scattered patches joint defence of those patches by related females, coupled with male emigration, should give rise to large, generally promiscuous multifemale/multimale groups. For species in which similarly stable groups of female relatives defend smaller patches, polygyny will prevail if groups follow the economic policy of allowing only one male to associate with them (Wrangham, 1980).   Cooperation among female kin has long been recognised as important to behaviour in many primate species, and Wrangham's model of female-bonded (FB) species has been influential in part because of its broad applicability.   Wrangham clearly distinguishes between environment and behaviour as selective forces influencing social organisation, and his ecological model differs from earlier correlational schemes in that it "... proceeds from a general assertion about mammalian biology to a set of derived and more specific propositions about how male and female primates distribute themselves," rather than from specific correlations to general explanations (Richard, 1985a).

PROXIMATE CONDITIONS OF GENE DISPERSAL IN NON-HUMAN PRIMATES

1. *Emigration as mechanism of inbreeding avoidance*

In primates, genes are dispersed primarily but not solely by emigration of individuals from their natal group. If the group is viewed as the basic unit for modelling social organisation and mating structure, there will be a tendency to equate migration of genes with migration of individuals, a tendency reinforced in selfish gene models by subordination of the carrier individual to the gene. Since gene dispersal is simply the opposite of inbreeding, the assumption that dispersal in itself may have evolved as a mechanism for avoiding inbreeding seems too tautological to be useful as a hypothesis guiding research. In any event, that hypothesis cannot be expected to predict migration (cf. Cheney & Seyfarth, 1983). The relation between dispersal and inbreeding avoidance has been discussed in detail by Moore and Ali (1984). Because gene dispersal need not require migration of individuals, and because individual migration may serve other functions than dispersal of genes, the accuracy of the prediction in any given instance might be spurious. What needs to be addressed are questions concerning the circumstances likely to favour one or another dispersal mechanism, or, where individual migration *is* at issue, what costs and benefits influence migrants' decisions as to which group to join (Cheney & Seyfarth, 1983).

2. *Male emigration*

In most species of primates, individual migration *is* an important mechanism of gene dispersal, and the primary agents of gene dispersal are males. Male emigration may result from competition for access to females, and the "reason" for a male's emigration seems clearest when he is observed to depart immediately after fighting with other males or following interactions with estrous females in a neighbouring group. However, access to females and success in mating depends on more than fighting ability, and fighting ability itself is not a simple matter. A male's relationships with females are governed by multiple and complex factors, including ties of kinship and early association, dominance relations with natal group females, and idiosyncratic features of individual-individual attraction. Consideration of the proximate circumstances of male migration must take these matters into account.

3. *Female emigration and female choice of mates*

Regular female emigration from the natal group has been reported for only a few species (Harcourt et al, 1976; Lancaster, 1984). However, it is by no means unusual for females to transfer from one group to another even in female-bonded species (Packer, 1979; Sugiyama & Ohsawa, 1982b; Berard, 1985). Recent studies have done much to correct past under-reporting of female migration and inattention to female choice in mating. O'Donald cautions against over-correction:

> "When females have no prior preference and exercise their choice by responding more readily to the more active and persistent of the males competing for their attention, differences in male behaviour alone determine the operation of sexual selection" (O'Donald, 1983, pp. 55-56).

But the old bias probably is reflected in any assumption of "no prior preference". There is room for exercise of female choice of partner at every stage of reproduction (Quiatt & Everett, 1981), and all the evidence suggests that it is more heuristic to assume that females exercise choice than that they do not.

The exercise of mating choice by either sex influences the behaviour of members of the opposite sex. Receptivity, proceptivity, and attractiveness appear to vary dynamically among individuals of both sexes and, over time, within individuals. Goy and Goldfoot found that sexual response by castrate rhesus monkeys varied with "... availability of partners that permit, encourage, or stimulate the male to display his maximum expression"; differences in performance among pairs could be "... conceptualised along some dimension such as sexual compatibility" (Goy & Goldfoot, 1975). Nadler discussed differential responsiveness of gorillas to one another in similar terms: "... some factor other than hormonal, presumably a higher cognitive variable, exerted an important influence on the sexual behaviour of these animals" (Nadler, 1975), and the observations of Miller and Quiatt, of two female gorillas paired with a series of different males over a three-year period (Quiatt et al, 1986; Miller, 1986) tend toward the same conclusion. There is, of course, no reason to suppose that it will be easier to say what makes for compatibility of potential mating partners in non-human than in human primates; the point is that where the emphasis is placed primarily on one sex, e.g. on male dominance and male choice, the dynamic interplay between individuals is likely to be overlooked.

4.   *Within-sex competition*

In Wrangham's model, species characterised by multi-female groups comprise two classes, female-bonded (FB) and non-female-bonded (Non-FB), and, as noted above, he accounts for the evolution of FB groups by evoking ecological pressures associated with high-quality patches of food in limited distribution. Female competition is strongly evident in FB species, and so is cooperation among subgroups who act together to supplant others from feeding sites (Wrangham, 1980). In FB species, social rank of individual females and of matrilineal subgroupings tends to be stable over long periods.   The idea that social rank confers access to high quality foods and improves nutrition "... is deeply entrenched in theories of the evolution of social behaviour....." (Richard, 1985a), and a number of studies of within-group competition and cooperation among females have focused mainly on feeding behaviour and infant handling (Small, 1984).   Females compete for mates as for other resources (Wasser, 1983; Crockett, 1984;   Whitten, 1984), and it would be interesting to know the extent to which mating choice in non-human primates may be influenced by corporate decision-making as is so frequently the case in human societies.   The question of cooperation here seems particularly relevant to consideration of population structure of FB species.   That issue apart, female-female competition is currently of keen interest to field primatologists.   The passive female whose fate rested on male competition during "receptive" phases of the reproductive cycle is a creature of past imagination;   today, most primatologists probably would agree with Hrdy and Williams that "... a female's competitive status relative to other females may be the single most pervasive influence on her reproductive success....." (Hrdy & Williams, 1983).

In most species of primates, males are the primary agents of gene dispersal, and fighting characteristically marks the transfer of a male from one group to another.   But the social behaviour of males is as complex as that of females, and the forms of male competition are similarly various and complex. They include sperm competition, in which what goes on at the haploid level may be masked by diploid tolerance of seeming promiscuity (Short, 1980), "stealing" matings while more dominant males are fighting over priority (Popp & DeVore, 1979), influencing a potential partner by exercising persuasive art or providing a food reward (Kuroda, 1984), rape, and of course fighting and conquest.

5.   *Integration of male and female dominance systems*

It is not simply the case that females compete for nutritional resources while males compete for females.   A virtue of Wrangham's model, as in any model which emphasises the working out of female dominance relations *primarily* in competition for nutritional resources and of male dominance relations *primarily* in competition for females, is that it raises the question how these two behavioural systems may be integrated.

Chapais, assessing the explanatory power of male dominance for mate choice in rhesus monkeys, found that the variance in male reproductive activity could not be explained entirely in terms of male rank:

"Female choice appeared to provide a pay-off asymmetry which could not be easily overcome ...   This suggests that the observed positive correlation between male rank and reproductive performance resulted from male-male competitive interactions *acting concurrently* with the capacity of males to influence female choice (e.g. through interferences in the consortships of lower-ranking males)" (Chapais, 1983).

Feedback must be important then, and, to a male attempting to influence reproductive choice, the efficacy of harassing competitors would seem to be enhanced by subsequent maintenance of close spatial association and synchronisation of behaviour with that of the prospective mate.

Although mount intrusions customarily have been assessed from the standpoint of male-male competition, Huffman has suggested that they may reveal as much about female control of reproductive activity (Huffman, 1984). Too, females may interrupt other females' copulations, and it sometimes happens that females engage in gang harassment of males (Quiatt, unpublished data from studies on Cayo Santiago; see also DeWaal, 1982). Such observations raise again the question whether or in what circumstances individual mating choice may reflect corporate decision-making. Where transient males work out dominance relations in dynamic interaction with a stable female structure, it should pay a male to establish close relations (reproductive or not) with a dominant matriline. On the whole, mount intrusions appear to constitute a set of behaviour too varied and too complex to be fitted into the narrow interpretive framework of direct male-male competition for access to females.

It is difficult to assess the influence of dominance rank on mating success (Fedigan, 1983), difficult also to establish rules of precedence by sex for the exercise of mating choice. Males and females can be ordered along a single ascending line of dominance, but one would be hard put to say what such a one-dimensional depiction of social order might mean. It seems clear that what is involved are two distinct systems (Meaney & Stewart, 1985), not one structure broken into two for convenience of analysis. How they may be integrated has important implications for male emigration and gene dispersal.

6.   *Dominance relations across sex as a factor in male emigration*

While they remain in their natal troop, male offspring of dominant mothers in FB species are dominant over sons of lesser dominant females, and they emigrate at a later age. Throughout early development they interact with a greater number of associates, especially with females (Quiatt, 1966; Berman, 1982), and tend "... to groom and to copulate with adult females who rank ... lower than their mothers....." (Cheney, 1978). For high-ranking young males, reproductive opportunities within the natal group appear to be greater than for low-ranking age peers, but they are reduced by inbreeding inhibition, and, to the extent that there exists a positive correlation between rank and relative size of matrilines, mating inhibition here works *against* dominant young males. Thus, dominance and uterine group association appear to exercise separate but complementary influences on the mating patterns of a number of primate species, insuring emigration of young males from their natal group.

7.   *Life history and demographic structure*

In many species of primates, mothers tend to promote early independence in males while tolerating more prolonged close contact with females. Meaney and Stewart (1985) relate differences in social play to these differences in maternal behaviour toward male and female offspring: differences in later social development and in extra-group activities reflect important between-sex differences in early life history. Studies of species-characteristic life histories usually have focused on males, who in many species may spend periods of their adult life in relative isolation or in company only with other males (Sugiyama, 1976); currently, more attention is directed toward females (e.g. Hasegawa & Hiraiwa-Hasegawa, 1983). A major difficulty in any assessment of individual differences in obtaining food and mates is separating effects of age and sex from those of social status (Richard, 1985a), and delineation of life history features clearly is basic to the task.

Demographic structure, social organisation, and mating patterns are closely interconnected. Age structure and sex ratio of a population play an important part in determining who an individual will interact with and who will be available as a mating partner (Dunbar, 1979). That group size influences mating pattern is evident in the correlation noted between opportunistic mating in a large group of wild chimpanzees (106 members) and restrictive mating in a smaller group (19 members). In the smaller group, only one female was likely to be in estrus at a given time, "... and the alpha male normally managed to monopolise her" (Hasegawa & Hiraiwa-Hasegawa, 1983).   Longitudinal demographic studies covering several generations have been carried out for a number of primate species, mostly familiar species of old world monkeys and the African apes.   They provide invaluable information concerning the effect of changes in group size and population density, and associated changes in available food and feeding circumstances, on reproductive and general social behaviour. Most such studies have been conducted in conjunction with provisioning. Where animals are provisioned, clumping of resources along with high population density consequent on improved nutrition may increase aggressive competition for food and significantly alter social and reproductive patterns. It has been suggested that in nature behaviour is less likely to provide clues to genealogy (Sugiyama &

Ohsawa, 1982a, cited by Richard, 1985a), while Altmann and Altmann (1979) have maintained that the matrilineal effect repeatedly observed in FB species in provisioned circumstances is much reduced in nature. A problem with any such comparison, given the short time-frame in which it is drawn, is deciding how representative conditions in *either* of the worlds compared may be of ecological circumstances in past time.

8.   *Can elemental socio-spatial units of population structure be delineated?*

The flow of genes within and across groups appears to be structured primarily by inhibition of mating with kin and other close associates.   The assumption that non-human primates recognise individual associates and classify them behaviourally as at least "kin" and "non-kin" has been implicit in most of the literature on species characterised by matrilineal organisation. Primatological intuition here is confirmed by recent experiment:

> Within groups, experiments suggest that animals may proceed
> beyond simple discriminations of kin and non-kin to create a
> taxonomy in which group members are both distinguished as
> individuals and grouped into higher order units, apparently on the
> basis of matrilineal kinship.   Across groups, observation indicates
> that male transfer is non-random....." (Cheney & Seyfarth, 1982.
> See also Cheney & Seyfarth, 1983).

Cheney and Seyfarth conclude that "... the social organisation of vervet monkeys is best regarded as a "community" of groups, within which individuals recognise each other and share a high degree of genetic relatedness...." (Cheney & Seyfarth, 1982).   That conclusion is similar to one arrived at by the ethnologist Delmos Jones concerning the larger organisation of human villages (Jones, 1968).

In multi-clan Hamadryas baboon groups (clan = group of closely associated males, each with a harem of females and bachelor followers), within-group gene flow is effected by female transfers between clans.   A juvenile female is likely to remain in the natal group when her mother transfers, but if she accompanies her mother, or when she transfers independently, she is not likely to return to become a member of her father's harem (Sigg et al, 1982).   Interestingly, females lost by a male to different rivals tend to reassemble in the same one-male unit later).

Gene flow in Sifaka populations appears not to depend on emigration and immigration of individuals.   "Spatial boundaries of groups do not coincide with social or reproductive boundaries even within a single breeding season" (Richard, 1985b).   Males mate with females of other groups in what appears to be a context of relationships maintained by continuity of individual recognition across groups.   The combination of kin-structured interactions and lack of strict territoriality in *Galago crassicaudatus* provides a similar contrast with earlier characterisations of prosimian sociality (Clark, 1986). Such observations point out the problematic character of the group; as far as local genetic structure of a population is concerned, what matters is the character of individual relationships within and across groups. The problem is partly but not wholly one of fuzzy boundaries and differential permeability. The Gelada baboon herd as reported by Ohsawa and Kawai is "... an ecological and social unit with fixed members in relation to domicile, even though the fringe is obscure and *may be continuous to the neighbouring herds*..." (Ohsawa & Kawai, 1975, emphasis added). They studied three herds - E, K and F, each composed of several one-male units, and two of these units shifted back and forth between E and K herds, spending 63.5 days (35% of total) in E, 67.5 days (37) in K, and 17.5 days (9) in both. Defining a boundary between groups is a problem for human observers but not for the animals, some of whom appear to maintain continuous membership in two groups.

In the past, as Richard has noted, "... studies of primate social behaviour emphasised interactions between members of a single group and categorised interactions between groups separately." But individuals who transfer from the natal group into adjacent or nearby groups are likely to encounter old associates, members of groups which have fissioned retain acquaintances across groups, and familiarity is created in any event by repeated encounters between groups, all of which suggests to primatologists today "... that intergroup relations are usefully studied as an integral part of social behaviour" (Richard, 1985a).

RECOGNITION OF INDIVIDUALS AND CLASSES OF INDIVIDUALS

Current mating system theory resembles other selfish gene explanations of social behaviour in that animals engaged in competition and cooperation with conspecifics are assumed to distinguish at some level of behaviour kin from non-kin. In primates, the close tie between mother and offspring produces a ramifying "matrilineal effect" on the distribution of social interactions. Early patterns of spatial and interactional association are continued in later life; thus, for females especially, and particularly in female-bonded species, social interactions appear to be predominantly kin-structured (Gouzoules, 1984).

That non-human primates respond to one another as individuals is evident. Cheney and Seyfarth (1982) analysed vervet monkeys' reactions to playbacks of vocalisations emitted on contact with individuals of various classes. Gouzoules and Gouzoules, in a similar study (of free-ranging rhesus monkeys), observed that when an infant's alarm vocalisation was recorded and played back in appropriate circumstances the mother looked in the direction of the concealed speaker while adult female associates, presumably more interested in what her reaction might be, looked at her (Gouzoules et al, 1984).

FAMILIARITY OVERRIDES GENEALOGY IN RECOGNITION OF "KIN"

The most thorough study of kin-correlated behaviour in mammals to date is that of Sherman and his associates, who found that while there were no consistent spatial differences among nine categories of Belding's ground squirrel female kin, only three consistently coexisted (i.e. in the literal sense of being alive simultaneously): mothers and daughters, littermate sisters, and non-littermate (half-) sisters.   These also, unlike the others, cooperated in proportion to relatedness (Sherman, 1981).

Interaction rates provide a better measure of familiarity than do "spatial differences", and these are likely to vary according to degree of relationship. Bekoff has stressed the importance of "facilitative" social environments over "determinative" effects of relatedness among siblings:

> "... [A] sibling is regarded as a sibling by virtue of early social experience and exposure during ... [development].   Thus, if genetically unrelated, or distantly related, individuals grow up under the same environmental conditions that characterise a species-typical sibling environment, ... they will behave as siblings because of a shared past history of repeated exposure and consequent familiarity with one another" (Bekoff, 1981).

A free-ranging rhesus monkey's circle of day-to-day associates usually is defined by continuing interaction and spatial association at a level of inclusiveness somewhere between family and matriline, and may include non-kin (Quiatt, 1986). Wu et al (1980), in a laboratory study of kin preference in infant *Macaca nemestrina*, obtained data supporting the idea of phenotype matching, i.e. their subjects appeared to "recognise" kin without previous experience. However, the sample was small (n=16), and Frederickson and Sackett were unable to replicate those findings with a larger sample of juvenile and young adult monkeys of the same species.   Their data suggested "... that familiarity was the whole basis for preferential choice among ... subjects.   Kinship neither influenced preferences as an independent factor nor added to familiarity in affecting social choices" (Frederickson & Sackett, 1984).

Gouzoules, in a comprehensive review of kin-correlated behaviour in primates, concludes that "As in Belding's ground squirrels, there appear to be limits to primate nepotism and the recognition of kin", and that "if the distribution of cooperative behaviour is contingent only upon familiarity [as would seem to be the case if the sole mechanism of kin recognition were prior association], it cannot be assumed that individuals are actually discriminating on the basis of the degree of genetic relatedness" (Gouzoules, 1984). This, of course, is a crucial point; resolution of the problem to which it refers is essential to productive application of kin altruism and mating system theory to evolutionary explanations of behaviour. If analysis cannot get beyond *behavioral* recognition of kin, the gap between ultimate and proximate explanations of social behaviour may remain insuperable.

In any event, the conclusion to be drawn from these studies is that distribution of social behaviour in primate societies appears to be governed by small interactive systems of spatial associates defined by mutual recognition in articulation with their habitat and the larger socio-ecological setting; groupings conventionally employed for description and comparison (e.g. family, matriline, the larger group) may be inappropriate referent units for analysis of "kin" recognition in nature (Quiatt, 1985).

DISCUSSION AND CONCLUSIONS

1.   The focus of data collection and data analysis in field studies of non-human primate behaviour, and of theory concerning that behaviour, has shifted from the group to the individual. The theory is characterised by specific assumptions concerning the adaptive significance of individual-individual competition within- and between-sex. Recent studies, many of them bearing on that competition, suggest that:

(a)   a number of species throughout the order are characterised by somewhat different dispersal and mating patterns than earlier had been supposed (Itani, 1980; Mehlman, 1981; Huffman, 1984; Terborgh & Goldizen, 1985; Clark, 1986; Richard, 1985b).

(b)   there is, in general, greater variation in mating pattern within species than has been assumed.

(c)   in some species, systematic mating preferences operate across as well as within groups (Richard, 1985b). In few or no species do matings within groups occur at random.

(d)   individual mating choices and associated cost-benefit decisions relating to reproduction are made in a context of wider social engagement. Other individuals and groups of individuals influence decisions directly and indirectly. It may be useful, in some circumstances, to view the exercise of mating choice as a corporate function.

2.   Field experiments indicating that non-human primates make finer class discriminations than simply kin vs. non-kin have employed audio playback

techniques which raise the question whether discriminative vocalisations might not be treated from a cognitive perspective as displaying semantic differences (Seyfarth et al, 1980a,b; Seyfarth, 1984; Gouzoules et al, 1984). But, however human observers may treat the cues by which they recognise that class discriminations are being made, in every species of animal (including our own) in which individuals recognise one another as kin or non-kin, recognition is evidenced in behaviour. Especially in the absence of a metalanguage by which we might converse with primates of another species about presumed acts of cognition, there is little analytic advantage in setting up parallel classes, rules, or systems of cognitive vs. behavioural kin recognition (but see Griffin, 1976, 1984, and Jolly, 1985).

The foregoing may seem obvious;   certainly it follows from Lloyd Morgan's tenet that no action be interpreted "... as the outcome of the exercise of a higher psychical faculty, if it can be interpreted as the outcome of one which stands lower in the psychological scale" (Morgan, 1894, cited by McFarland, 1985).   However, my object in emphasising that kin are recognised by overt behaviour is not to promote behaviourism but to note implications for anthropological comparison of population structure in human and non-human primate species. It is customary in anthropology to contrast "conventional" (i.e. cultural) systems of human kinship with "natural" systems of non-human animal kinship, e.g. the mother-offspring units of certain prosimians, the two-parent families of gibbons and siamangs, the extended matrilineal groupings of rhesus monkeys and Japanese macaques, etc. What is different about human life, it has seemed, are more or less formal procedures such as adoption and initiation which allow us to treat an individual "as if" he or she were a member of some natural class of associates.   But if we allow that kin recognition is in all species the formal outcome of habit the difference vanishes (which is not to say that differences in behaviour by which animals of different species recognise kin are of no account). The point is that kin recognition by non-human primates, though complex and problematic as a subject for research, is observable only when expressed in behaviour. In that sense, and in the sense that it is learned (as appears to be the case), it is conventional as in our own species.

3.   In many species of primates, populations are structured locally by networks of close associates and familiar individuals, and the boundaries of those networks may not be coterminous with those of group, herd, band, village, etc. In this respect the mating patterns of non-human primates appear to resemble our own far more than most of the literature on behaviour would lead one to suppose; and, in this respect, to the extent that there is congruence within species between maps of genetic structure, social organisation, and individual

behaviour, there is congruence across species, too. The problem of course is to move from sociobiological theories of behaviour in evolution to understanding what particular similarities and differences within and across species reveal about adaptation. That move is less likely to be advanced by further amplification of theory than by testing current theory against the proximate contexts of real behaviour.

4. Current mating system theory utilises logically coherent models of ultimate causation, which can be judged as compatible or incompatible with observed behaviour, but which cannot be applied with confidence to reconstruction of events in the evolution of that behaviour. Such models are of inestimable value in ordering our thoughts about the genetic logic of behaviour in evolution, but it is important not to lose sight of the *behaviour* and to keep in mind that "... understanding of the selective forces ... [which shape a] ... social system must rest on a detailed and accurate knowledge of the proximate forces that maintain it" (Brockelman & Srikosamatara, 1984). This is especially important where patterns of mating and gene dispersal are concerned, for explanatory models which perpetuate old notions of the group as *the* natural unit of analysis may lead to overlooking or misconstruing spatial and interactional data which bear on subgroup or intergroup processes. Also, because individual cost-benefit decision-making functions are central to mating system models, it is necessary to ask whether a meaningful distinction can be made between the processes involved in those and the very similar processes underlying models of learning. The question is not a trivial one, and it is especially important when what is at issue is the behaviour of nonhuman primates, for many species of which, as is increasingly clear, learning and (presumably) cognition must operate in much the same way as in ourselves. I think the answer to this question will have significant implications for practical research, for if we are to understand the behaviour which regulates gene flow and structures populations it will not do to confuse or conflate theories of learning and theories of evolutionary adaptation.

REFERENCES

Altmann, S.A. & Altmann, J. (1979). Demographic constraints on behavior and social organization, In: I. S. Bernstein & E. O. Smith (eds.), Primate Ecology and Human Origins, pp. 47-62. New York: Garland.

Bekoff, M. (1981). Mammalian sibling interactions, In: D. J. Gubernick & P. H. Klopfer (eds.), Parental Care in Mammals, pp. 307-346. New York: Plenum.

Berard, J. (1985). Female transfer in free-ranging rhesus monkeys. Paper presented at the Annual Meeting of the American Society of Primatologists, Niagara Falls, New York.

Berman, C.M. (1982).   The ontogeny of social relationships with group companions among free-ranging infant rhesus monkeys.   I. Social networks and differentiation. Animal Behaviour, 30, 149-162.

Brockelman, W.Y. & Srikosamatara, S. (1984).   Maintenance and evolution of social structure in gibbons,   In: H. Preuschoft, D. L. Chivers, W. Y. Brockelman & N. Creel (eds.), The Lesser Apes, pp. 298-323. Edinburgh University Press.

Carpenter, C.R. (1965). The howlers of Barro Colorado, In: I. DeVore (ed.), Primate Behavior, Field Studies of Monkeys and Apes, pp. 250-291. New York: Holt, Rinehart & Winston.

Chalmers, N. (1979).   Social Behaviour in Primates.   Baltimore: University Park Press.

Chapais, B. (1983).   Reproductive activity in relation to male dominance and the likelihood of ovulation in rhesus monkeys.   Behavioral Ecology & Sociobiology, 12), 215-228.

Cheney, D.L. (1978).   Interactions of immature male and female baboons with adult females. Animal Behaviour, 26, 389-408.

Cheney, D.L. & Seyfarth, R.M. (1982).   Recognition of individuals within and between groups of free-ranging vervet monkeys. American Zoologist, 22, 519-529.

Cheney, D.L. & Seyfarth, R.M. (1983).   Non-random dispersal in free-ranging vervet monkeys: social and genetic consequences. American Naturalist, 122, 392-412.

Clark, A.B. (1986).   Sociality in a noctural "solitary" prosimian: Galago crassicaudatus. International Journal of Primatology, 6, 581-600.

Crockett, C.M. (1984).   Emigration by female red howler monkeys and the case for female competition,   In: M. F. Small (ed.),   Female Primates: Studies by Women Primatologists, pp. 159-173. New York: Liss.

DeVore, I. (ed.) (1965).   Primate Behavior, Field Studies of Monkeys and Apes. New York: Holt, Rinehart & Winston.

DeWaal, F. (1982).   Chimpanzee Politics: Power and Sex among Apes. New York: Harper & Row.

Dunbar, R.I.M. (1979).   Population demography, social organization, and mating strategies, In: I. S. Bernstein & E. O. Smith (eds.), Primate Ecology and Human Origins, pp. 67-68. New York: Garland.

Fedigan, L. M. (1983).   Dominance and reproductive success in primates. Yearbook of Physical Anthroplogy, 26, 91-129.

Fredrickson, W.T. & Sackett, G.P. (1984). Kin preferences in primates (Macaca nemestrina): relatedness or familiarity?   Journal of Comparative Psychology, 98, 29-34.

Gilmore, H.A. (1981).   From Radcliffe-Brown to sociobiology: some aspects of the rise of primatology within physical anthropology. American Journal of Physical Anthropology, 56, 387-392.

Gouzoules, S. (1984). Primate mating systems, kin associations, and cooperative behavior: evidence for kin recognition?   Yearbook of Physical Anthropology, 27, 99-134.

Gouzoules, S., Gouzoules, H. & Marler, P. (1984).   Rhesus monkey (Macaca mulatta) scream vocalizations: representational signaling in the recruitment of agonistic aid. Animal Behaviour, 32, 182-193.

Goy, R.W. & Goldfoot, D.A. (1975). Neuroendocrinology: animal models and problems of human sexuality. Archives of Sexual Behaviour, **4**(4), 405-420.

Greenwood, P.J. (1980). Mating systems, philopatry, and dispersal in birds and mammals. Animal Behaviour, **28**, 1140-1162.

Griffin, D.R. (1976). The Question of Animal Awareness: Evolutionary Continuity of Mental Experience. New York: Rockefeller University Press.

Griffin, D.R. (1984). Animal Thinking. Cambridge, Massachusetts: Harvard University Press.

Hall, K.R.L. & DeVore, I. (1965). Baboon social behaviour, In: I. DeVore (ed.), Primate Behavior, Field Studies of Monkeys and Apes, pp. 53-110. New York: Holt, Rinehart & Winston.

Harcourt, A.H., Steward, K.S. & Fossey, D. (1976). Male emigration and female transfer in wild mountain gorilla. Nature, **263**, 226-227.

Hasegawa, T. & Hiraiwa-Hasegawa, M. (1983). Opportunistic and restrictive matings among wild chimpanzees in the Mahale Mountains, Tanzania. Journal of Ethology, **1**, 75-85.

Hrdy, S.B. & Williams, G.C. (1983). Behavioral biology and the double standard, In: S. K. Wasser (ed.), Social Behavior of Female Vertebrates, pp. 3-17. New York: Academic Press.

Huffman, M.A. (1984). Mount-sequence intrusion and male-male competition in *Macaca fuscata*. International Journal of Primatology, **5**, 349 (Abstr.).

Itani, J. (1980). Social structures of African great apes. Journal of Reproduction & Fertility, Supplement, **28**, 33-41.

Jay, P. (1965). The common langur of north India, I: I. Devore (ed.), Primate Behavior, Field Studies of Monkeys and Apes, pp. 197-249. New York: Holt, Rinehart & Winston.

Jolly, A. (1985). The evolution of primate behavior. American Scientist, **73**, 230-239.

Jolly, C.J. & Plog, F. (1979). Physical Anthropology and Archaeology, 2nd ed. New York: Alfred A. Knopf.

Jones, D.J. (1968). The multivillage community: village segmentation and coalescence in northern Thailand. Behavioral Science Notes, **3**, 149-174.

Koford, C.B. (1965). Population dynamics of rhesus monkeys in Cayo Santiago, In: I. DeVore (ed.), Primate Behavior, Field Studies of Monkeys and Apes, pp. 160-174. New York: Holt, Rinehart & Winston.

Kuroda, S. (1984). Socio-sexual behavior of the pygmy chimpanzee. International Journal of Primatology, **5**, 399 (Abstract of film).

Lack, D. (1968). Ecological Adaptations for Breeding in Birds. London: Methuen.

Lancaster, J.B. (1984). Introduction, In: M. F. Small (ed.), Female Primates: Studies by Women Primatologists, pp. 1-10. New York: Liss.

McFarland, D. (1985). Animal Behavior: Psychobiology, Ethology, and Evolution. Menlo Park: Benjamin/Cummings.

Meaney, M.J. & Stewart, J. (1985). Sex differences in social play: the socialization of sex roles. Advances in the Study of Behavior, **15**, 1-58.

Mehlman, P. (1986).   Male intergroup mobility in a wild population of the barbary macaque (*Macaca sylvanus*), Ghomaran Rif Mountains, Morocco. American Journal of Primatology, **10**, 67-81.

Miller, L. (1986).   Improving fertility in captive gorillas.   M.A. thesis, University of Colorado, Denver.

Moore, J. & Ali, R. (1984).   Are dispersal and inbreeding avoidance related? Animal Behaviour, **32**, 94-112.

Morgan, C.L. (1894). Introduction to Comparative Psychology. London: Scott.

Nadler, R.D. (1975). Sexual cyclicity in captive lowland gorillas. Science, **189**, 813-814.

O'Donald, P. (1983).   Sexual selection by female choice,   In: P. Bateson (ed.), Mate Choice, pp. 53-66. Cambridge University Press.

Ohsawa, H. & Kawai, M. (1975).   Social structure of Gelada baboons,   In: Contemporary Primatology, Papers of the 5th International Congress of Primatologists, Nagoya, 1974, pp. 464-469. Basel: Karger.

Packer, C. (1979). Intertroop transfer and inbreeding avoidance in *Papio anubis*. Animal Behaviour, **27**, 1-36.

Popp, J.L. & DeVore, I. (1979).   Aggressive competition and social dominance theory: synopsis. In: D. A. Hamburg & E. R. McCown (eds.),   The Great Apes, pp. 317-338. Menlo Park: Benjamin/Cummings.

Quiatt, D. (1966).   Social dynamics of rhesus monkey groups.   Ph.D. dissertation. Ann Arbor: University Microfilms.

Quiatt, D. (1985).   The "household" in non-human primate evolution: a basic linking conception. Anthropologea Contemporanea, **3**, 187-193.

Quiatt, D. (1986).   Juvenile/adolescent role functions in a rhesus monkey troop: an application of household analysis to non-human primate social organization,   In: J. Else & P. Lee (eds.),   Primate Ontogeny, Cognition and Social Behaviour. Cambridge University Press.

Quiatt, D. & Everett, J. (1982).   How can sperm competition work?   American Journal of Primatology, Supplement 1, 161-169.

Quiatt, D., Miller, L. & Cambre, R. (1986).   Overt behavior correlates of menstruation and ovulation in a lowland gorilla,   In: D. M. Taub & F. A. King (eds.),   Current Perspectives in Primate Biology, pp. 32-41. New York: Van Nostrand Reinhold.

Richard, A. (1985a). Primates in Nature. New York: Freeman.

Richard, A. (1985b).   Social boundaries in a Malagasy prosimian, the sifaka (*Propithecus verreauxi*). International Journal of Primatology, **6**, 553-568.

Selander, R.K. (1972).   Sexual selection and dimorphism in birds.   In: B. Campbell (ed.), Sexual Selection and the Descent of Man, pp. 180-230. Chicago: Aldine.

Seyfarth, R.M. (1984).   What the vocalizations of monkeys mean to humans and what they mean to the monkeys themselves,   In: R. Harre & V. Reynolds (eds.),   The Meaning of Primate Signals,   pp. 43-56.   Cambridge University Press.

Seyfarth, R.M., Cheney, D.L. & Marler, P. (1980a). Monkey responses to three different alarm calls: evidence for semantic communication and predator classification. Science, **210**, 801-803.

Seyfarth, R.M., Cheney, D.L. & Marler, P. (1980b). Vervet monkey alarm calls: semantic communication in a free-ranging primate. Animal Behaviour, **28**, 1070-1094.

Sherman, P.W. (1981). Kinship, demography, and Belding's ground squirrel nepotism. Behavioral Ecology & Sociobiology, **8**, 251-260.

Short, R.V. (1980). The origins of human sexuality, In: C. R. Austin & R. V. Short (eds.), Reproduction in Mammals, 8. Human Sexuality, pp. 1-33. Cambridge University Press.

Sigg, H., Stolba, A., Abbeglen, J.-J. & Dasser, V. (1982). Life history of Hamadryas baboons: physical development, infant mortality, reproductive parameters and family relationships. Primates, **23**, 473-487.

Small, M.F. (ed.) (1984). Female Primates: Studies by Women Primatologists. New York: Liss.

Sugiyama, Y. (1976). Life history of male Japanese monkeys. Advances in the Study of Behavior, **7**, 255-284.

Sugiyama, Y. & Ohsawa, H. (1982a). Population dynamics of Japanese monkeys with special references to the effect of artificial feeding. Folia Primatologica, **39**, 238-263.

Sugiyama, Y. & Ohsawa, H. (1982b). Population dynamics of Japanese macaques at Ryozenyama. III. Female desertion of the troop. Primates, **23**, 31-44.

Terborgh, J. & Goldizen, A.W. (1985). On the mating system of the cooperatively breeding saddle-backed Tamarin. Behavioral Ecology & Sociology, **16**, 293-299.

Wasser, S.K. (1983). Reproductive competition and cooperation among yellow baboons, In: S. K. Wasser (ed.), Social Behavior of Female Vertebrates, pp. 349-390. New York: Academic Press.

Whitten, P.L. (1984). Competition among female vervet monkeys, In: M. F. Small (ed.), Female Primates: Studies by Women Primatologists, pp. 127-140. New York: Liss.

Wittenberger, J.F. (1979). The evolution of mating systems in birds and mammals, In: P. Marler & J. G. Vandenbergh (eds.), Handbook of Behavioral Neurobiology, vol.3, Social Behavior and Communication, pp. 271-349. New York: Plenum.

Wittenberger, J.F. (1980). Group size and polygamy in social animals. American Naturalist, **115**, 197-222.

Wrangham, R.W. (1980). An ecological model of female-bonded primates groups. Behaviour, **75**, 262-300.

Wu, H.M.H., Holmes, W.G., Medina, S.R. & Sackett, G.P. (1980). Kin preference in infant

INBREEDING IN HUMAN POPULATIONS: AN ASSESSMENT OF THE COSTS

A. H. BITTLES[1] and E. MAKOV[2]

[1] *Department of Anatomy and Human Biology, King's College, University of London, U.K.*
[2] *Department of Statistics, University of Haifa, Israel*

INTRODUCTION

One of the earliest specific demonstrations of inbreeding effects in man was reported in 1902 when Garrod showed that of eighteen cases of the inborn error of metabolism alkaptonuria diagnosed in Europe and North America, twelve had been born to parents who were first cousins. However, in order to maintain a sense of perspective on the level of morbidity resulting from consanguinity, Garrod reminded his readership of the extreme rarity of the condition in first cousin offspring as a whole. In London, only six alkaptonurics had been detected among the estimated 50,000 first cousin progeny then resident in the city. Unfortunately this balanced approach to the effects of human inbreeding has not always been followed. As a result, the high rates of specific disorders identified in particular endogamous groups frequently have been inappropriately adopted by other workers as the bases for predictions on the probable effects of inbreeding in large human populations, or even in the entire species. It must be stressed that while inbreeding may be costly in families or groups known to carry recessive lethal or sub-lethal genes, this outcome is by no means inevitable. The aim of the present chapter is therefore to provide an overall assessment of the risks associated with various degrees of human inbreeding at the population, as opposed to the individual or group, level.

FACTORS INFLUENCING THE TIMING AND EFFECTS OF HUMAN INBREEDING DEPRESSION

Disorders which would be expected to show an increased incidence with inbreeding are those resulting from the expression in homozygotes of single, autosomal recessives or clusters of rare, polygenic recessives. The rarer the frequency of a deleterious recessive gene, the greater the probability that it will be expressed in the progeny of related persons rather than in the general non-consanguineous population.   For example, in Italy the incidence at birth of cystic fibrosis is approximately 1:2,000, whereas that of phenylketonuria is 1:12,000.   In keeping with theoretical predictions, while 4.5% of cystic fibrosis cases were found to be born to cousin marriages, the comparable figure for phenylketonuria was 11.2% (Romeo et al, 1981, 1983). Besides disorders with a definable, autosomal recessive mode of inheritance, it has been proposed that recessive genes also may be implicated in the etiology of chromosomal anomalies.   According to this hypothesis, the increased incidence of Down's syndrome cases observed in the offspring of certain first cousin unions may have arisen from mitotic non-disjunction in the homozygous fertilised ovum, due to the action of a recessive gene or genes (Alfi et al, 1980; Yokoyama et al, 1981). However, in such cases the precise nature of the proposed mechanism remains, at best, speculative.

Any assessment of disease states which may or may not be caused by the action of recessive genes, and hence be influenced by inbreeding, is complicated by disorders which simulate Mendelian inheritance (McKusick, 1978).   A number of examples associated with this phenomenon are given in Table 1.   In addition, there is a wide range of basic biological and social factors which may significantly influence the outcome of a pregnancy (Table 2).   Appropriate allowance for each must be made before attempting quantification of the risks accruing from human inbreeding.

Table 1:   The simulation of single gene disorders

Multifactorial disorders
Undetected chromosomal aberrations
Disorders of the maternal genotype
Congenital infections
Chemical teratogens

Table 2:   Inbreeding depression: potential distortional factors

Biologically distinct populations
Differential biological inferiority
Differential socioeconomic status
Polygamy/polygyny
Adoption
Non-paternity

Table 3: Timing of the expression of rare, recessive genes

Period	Generation
1. **Antenatal** Increased sterility Decreased fertility Increased fetal loss (abortions or stillbirths)	Parents
2. **Postnatal, pre-reproductive** Increased morbidity Increased mortality	Offspring
3. **Postnatal, reproductive** Increased morbidity Increased mortality Decreased fertility	Offspring

The timing and observable effects of recessive gene action are summarised in Table 3. The accuracy with which each parameter can be measured varies considerably, the greatest degree of uncertainty being associated with events during the antenatal period. By general agreement, inbred marriages have reduced levels of primary sterility when compared with non-consanguineous matings (reviewed in Bittles, 1980; MacCluer, 1980). This appears to contradict recent reports which suggest that there may be failure to initiate a pregnancy when the partners share HLA antigens, due to absence in the female of cytotoxic antibody against her partner's T and B lymphocytes (Mowbray et al, 1985). However, the results of these studies have themselves been queried (Adinolfi, 1986), rendering the entire situation uncertain. The relationship between inbreeding and fetal loss is equally unclear as a number of studies have suggested no excess loss or even reduced numbers of abortions with inbreeding (Böök, 1957; Stevenson & Warnock, 1959), possibly attributable to reduced risks of *erythroblastosis fetalis* (Stern & Charles, 1945) and/or pre-eclamptic toxaemia (Stevenson et al, 1971). Conversely, increased fetal loss also has been reported (Bigozzi et al, 1971) and in their study on Hirado, Schull and Neel (1972) assumed a 4% greater early fetal loss in first cousin offspring. Until an accurate evaluation of antenatal losses is available for non-consanguineous pregnancies, it is impossible to determine the additional effects of inbreeding, whether beneficial or detrimental. Statistical estimates of total antenatal losses have varied from 48% (James, 1975) to 78% (Roberts & Lowe, 1975) while biochemical monitoring using human chorionic gonadotropin (hCG) levels have indicated early post-implantation losses of between 20% (Whittaker et al, 1983) and 43% (Miller et al, 1980). With such discrepant baseline results, the analysis of inbreeding effects in the antenatal period virtually becomes guesswork.

Overall, the estimation of postnatal mortality and morbidity presents a somewhat lesser problem. As may be deduced from Tables 1 and 2, one of the major difficulties in the postnatal phase resides in an unequivocal demonstration that excess mortality and/or morbidity observed in consanguineous offspring was indeed caused by the expression of recessive genes. Since the expertise which would be required for such a demonstration often exceeds the resources available to investigators, especially in developing countries, suitably matched control groups are essential. In many studies on human inbreeding the non-consanguineous control groups employed have singularly failed to match their consanguineous counterparts, most commonly in terms of socioeconomic status. By definition, failure to compare adequately matched consanguineous and non-consanguineous groups will lead to bias in quantifying the effects of inbreeding, the extent of the bias being dependent both on the nature of the non-genetic variable under investigation and the timing of its maximal interaction with individuals' genotypes. Where consanguineous marriages have been favoured over many generations, allowance for an additional genetic variable, the effect(s) of inbreeding in the parental generation, also may be necessary (Schull et al, 1970).

THE ASSESSMENT OF HUMAN INBREEDING EFFECTS

Attempts at assessing the costs of human inbreeding have a long and often intriguing history. In 1875 George Darwin, himself the progeny of a first cousin marriage (see Macfarlane, this volume), published an account of his investigation into the relative fertility and mortality associated with first cousin unions. This was based on data gleaned from impeccable sources, *Burke's Landed Gentry* and *The English and Irish Peerage*. He reported that in terms of net fertility, as measured by the numbers of sons surviving infancy, first cousin marriages held a slight advantage over their non-consanguineous peers, having between 1.92 and 2.07 surviving sons as opposed to the average 1.91 male survivors of marriages between non-relatives (Darwin, 1875a). There was evidence of increased infant mortality with inbreeding, which Darwin estimated at between 18.1% and 21.1% in the first cousin offspring versus 18.2% in the non-consanguineous controls. However, his *pièce de résistance* was the demonstration of reduced fitness in the inbred group (Darwin, 1875b), by contrasting the lower proportion of first cousin offspring among Oxford and Cambridge "boating men", 2.4%, with the equivalent proportion among their non-sporting social peers, 3.0 to 3.5%.

While endowed with great charm, the results of Darwin's studies in themselves cannot be used to define the nature and extent of human inbreeding

Table 4: Mortality studies on first cousin offspring

Investigators	Study area	
1. Sutter & Tabah (1952, 1953)	France:	Loir et Cher
2. Sutter & Tabah (1952, 1953)	France:	Morbihan
3. Böök (1957)	Sweden:	Norrbotten
4. Schull (1958)	Japan:	Kure
5. Neel & Schull (1962)	Japan:	Hiroshima
6. Neel & Schull (1962)	Japan:	Nagasaki
7. Stevenson et al (1966)	Brazil:	Sao Paulo
8. Stevenson et al (1966)	India:	Bombay
9. Stevenson et al (1966)	Singapore	
10. Schull et al (1970)	Japan:	Hirado
11. Yamaguchi et al (1970)	Japan:	Fukuoka
12. Fraser & Biddle (1976)	Canada:	Quebec
13. Rao & Inbaraj (1977)	India:	Tamil Nadu (rural)
14. Rao & Inbaraj (1977)	India:	Tamil Nadu (urban)
15. Rao & Inbaraj (1979)	India:	Tamil Nadu (rural)
16. Rao & Inbaraj (1979)	India:	Tamil Nadu (urban)
17. Lindelius (1980)	Sweden	
18. Asha Bai et al (1981)	India:	Tamil Nadu
19. Shami & Zahida (1981)	Pakistan:	Lahore
20. Edo et al (1985)	Spain:	Zamora

depression, if only because of the highly restricted nature of the sample group. To obtain a global estimate of the effects of inbreeding, assuming that such a value would have real biological significance, it is necessary to consider more representative studies, a series of which have been listed in Table 4.   The studies were drawn both from populations in which outbreeding is the widely accepted norm, for example, Western Europe and North America, and from Japan and South India with their long traditions of consanguineous marriages. Surveys conducted on small population isolates were excluded because of the possible, conflicting effects of factors such as genetic drift.

The data, comparing the incidences of stillbirths and deaths in the first month and first year of life in non-consanguineous offspring, and in first cousin progeny with a coefficient of inbreeding (F) of 0.0625, are presented as scatter diagrams in Figures 1 to 3.   The three time periods chosen represent the largest and most reliable data sets available and in contemporary populations they account for approximately 70% to 80% of all pre-reproductive mortality (Neel & Schull, 1962; Macfarlane & Mugford, 1984).   The result from each study is marked as a point, the horizontal coordinate of which is the proportion of deaths in the non-consanguineous group, with deaths among first cousin offspring as the vertical coordinate.   In such a scatter diagram any systematic departure from the 45° line is indicative of an inbreeding effect: upward being associated with detrimental effect, downward suggesting an advantageous outcome.   The greater the strength of the inbreeding effect, the further the position of the point above or below the line.

The thirteen studies for which data on stillbirths were available are shown in Figure 1.   Since the data points are scattered quite evenly around the 45° line, the conclusion must be that minimal inbreeding effect is detectable and therefore deaths which occurred during this period were predominantly "environmental" in origin.   Quite different results were apparent when data from the eleven available studies on deaths in the first month of life (Figure 2) and the fifteen studies on deaths in the first year of life (Figure 3) were plotted.   With one exception, from South India, all points plotted above the 45° line indicating increased mortality in the first cousin group.   The mean excess mortality over the three periods totalled 3.3%; there were a mean 0.5% extra stillbirths and 1.4% and 2.8% excess deaths in the first month and first year of life respectively.   The studies from single national sources, for example Japan, tended to give similar results with respect to inbreeding effects.   This may indicate either national differences in the risks pertaining to inbreeding, possibly

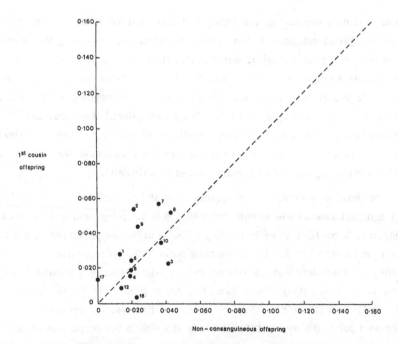

Figure 1:   Scatter diagram of the proportion of stillbirths in first cousin
               offspring versus non-consanguineous offspring.

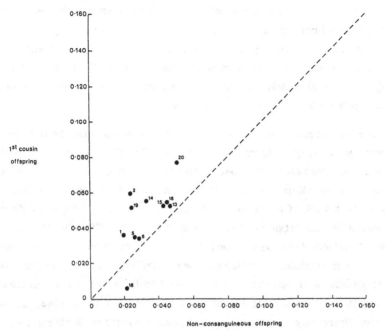

Figure 2: Scatter diagram of the proportion of first month deaths in first cousin offspring versus non-consanguineous offspring.

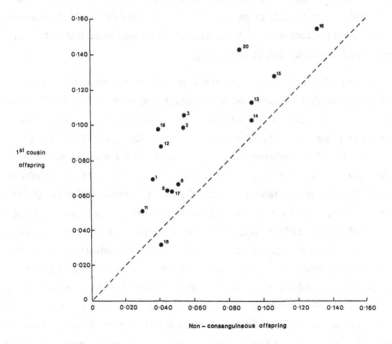

Figure 3: Scatter diagram of the proportion of first year deaths in first cousin offspring versus non-consanguineous offspring.

secondary to the history and extent of consanguineous marriages in the community, or merely reflect a greater similarity in the experimental design adopted by the various groups of investigators in each country.   The high mortality of the non-consanguineous groups in large-scale studies from Tamil Nadu (Rao & Inbaraj, 1977, 1979) clearly showed that non-genetic causes of death remained prevalent in South India during the study period.

The measurement of postnatal morbidity is less precise, as it is largely dependent on the diagnostic criteria adopted.   For this reason, very variable incidences of morbidity have been· reported, particularly in small surveys (reviewed in Freire-Maia & Elisbao, 1984).   In the United Kingdom, it has been estimated that 4-5% of all newborn infants have some form of genetic disorder or congenital malformation (Medical Research Council, 1978).   Generally, major congenital defects have been found to be more common in consanguineous offspring.   For example, their incidence was 0.4% higher at birth in the infants of first cousins in Hiroshima and Nagasaki (Neel, 1958) with a total excess morbidity to pre-reproductive age of 1.4-4.1% (Schull & Neel, 1965), although this latter figure may have been an overestimate caused by inadequate correction for socioeconomic differentials (Neel et al, 1970a).   In Tamil Nadu (Rao & Inbaraj, 1980), no significant excess of congenital defects was apparent at birth in first cousin (F = 0.0625) or in uncle/niece offspring (F = 0.125), however the high background effect of environmental variables may have interfered with the estimation of morbidity due to inbreeding.

Somewhat conflicting conclusions have been drawn with respect to the effect of inbreeding on tests of intellectual capacity (Böök, 1957;   Slatis & Hoene, 1961).   A mean 3-4% decline in the school performance of first cousin progeny was observed in Hiroshima and Nagasaki (Schull & Neel, 1965), but on Hirado no significant inbreeding effect on IQ scores was found (Neel et al, 1970b).   The discrepancy may be explained by the fact that in Hiroshima and Nagasaki there was a sizeable contribution to the consanguineous group from children born to parents who had migrated into the cities from outlying rural areas (Neel et al, 1970b), thus violating at least one of the pre-conditions relating to control groups listed in Table 2.   Nevertheless, a large detailed survey on Arab children in Israel did demonstrate lower mean IQ scores with inbreeding (Bashi, 1977).   In both mental ability and achievement tests the highest mean level of performance was attained by non-consanguineous children, and the lowest by double first cousin offspring (F = 0.125), although the difference in IQ scores was small and the double first cousin group was of below average socioeconomic status.   The most significant finding was the increased variation in IQ scores among double first cousin progeny indicating that, at least in certain individuals, decreased intellectual performance may have resulted from the expression of deleterious recessive genes.   This conclusion subsequently was supported by a similar study conducted in North India (Agrawal et al, 1984).

Hence the evidence from both mortality and morbidity studies indicates that in consanguineous offspring the expression of deleterious recessive genes occurs principally in the postnatal period, any effects in the antenatal period largely being obscured by interference from non-genetic factors. Characteristically, consanguineous marriages have a compensatory advantage over their non-consanguineous counterparts in terms of greater numbers of pregnancies and enhanced net fertility (Schull et al, 1962; Schull & Neel, 1972). This may result from reduced birth intervals operating to replace infants dying early in the postnatal period (Record & Armstrong, 1975;   Ericksen et al, 1979). Alternatively, in areas where consanguineous marriages have been favoured over many generations, social influences also may be involved (Schull et al, 1968). In either case, the major effect on the gene pool would be to retard significantly inbreeding-associated elimination of recessive mutants (Bittles et al, 1985).

THE CALCULATION OF LETHAL AND MORBID GENE EQUIVALENTS

In an attempt to translate the excess risks associated with inbreeding into numbers of lethal recessive genes carried by human populations, Morton et al (1956) suggested the model:

$$S = e^{-A-BF}$$

where S is the proportion of survivors in a population, A is deaths expressed under random mating, B is deaths arising from recessive gene expression via inbreeding, and F, the coefficient of inbreeding, is the fraction of gene loci homozygous as a result of consanguinity. By transforming S, A and B can be estimated by linear regression or, as subsequently was suggested, by weighted regression (Smith, 1967, 1969). Over the years a large array of values for B has been amassed, describing groups and populations both in terms of mortality as lethal gene equivalents (lethons), and morbidity as detrimental gene equivalents (morbons).   Representative general estimates on the numbers of lethal and detrimental gene equivalents in humans are given in Table 5. Manifestly, there is considerable variation between estimates with much lower values cited in the more recent studies.

Using the figure of 2.2 lethal equivalents to describe the genetic load carried by the average non-consanguineous individual (Cavalli-Sforza & Bodmer,

Table 5:   Mean number of lethal/detrimental gene equivalents per individual

Study	Period	Lethons	Morbons
Morton et al (1956)	Late fetal to early adult	3.5	
Morton (1960)	Late fetal to early adult		1.5
Cavalli-Sforza & Bodmer (1971)	Pre-reproductive	2.2	
Freire-Maia (1984)	Fetal to adult	1.5	0.5

1971), it was calculated that an excess 13% of first cousin progeny (F = 0.0625), 24% of uncle/niece offspring (F = 0.125) and 42% of incest offspring (F = 0.25) would be expected to die before reaching reproductive age (May, 1979). It is obvious that at first cousin level this estimated additional mortality is far in excess of the range of values shown in Figures 1 to 3, and the mortality predicted for uncle/niece offspring equally is significantly greater than the observed levels in Tamil Nadu (Rao & Inbaraj, 1977, 1979) and Karnataka (Bittles et al, 1985). It is only with respect to the progeny of incestuous matings that reported mortality levels (Carter, 1967; Baird & McGillivray, 1982) approach the calculated values. However, this may be due principally to an absence of non-consanguineous controls in these studies. Even when a sizeable control group was used (Seemanová, 1971), non-genetic causes including young maternal age were shown to be strongly implicated in the very high numbers of deaths and defects reported (Bittles, 1979).

Four possible reasons can be identified for the apparent overestimation of inbreeding effects by the linear regression method. First, as noted above, the data may be drawn from studies with inadequate control groups. The downward re-evaluation of lethal and detrimental gene equivalents with time (Table 5) appears to confirm this view, as does the lower net level of death and major defect reported in an incest study which incorporated a well-matched control group (Adams & Neel, 1967). Second, the implicit assumption behind the exponential model, that genetic and environmental causes of death are independent in their action (Morton et al, 1956), appears unwarranted. Third, there has been a tendency to disregard confidence limits and to concentrate only on mean values. As illustrated in Figure 4, based on the data of Sutter and Tabah (1952, 1953), 95% confidence limits for the proportion of survivors (dashed line) indicate that the model is not significant at α = 0.05 (Bittles & Makov, 1985), yet a lethal equivalent value is commonly quoted for this study.   Fourth, assuming that a statistically significant difference does exist between the consanguineous and non-consanguineous groups, the actual values obtained for lethal or detrimental gene equivalents will be dependent on the particular combination of model/transformation and error distribution selected.   A logarithmic transformation and normal distribution originally were chosen by Morton et al (1956) but no *a priori* reason for their exclusive adoption is apparent.   It may be that one or more alternative transformation/error distribution combinations would provide more representative mortality and morbidity estimates but this is difficult to prove or disprove given the relative paucity of human inbreeding data (Makov & Bittles, 1986).

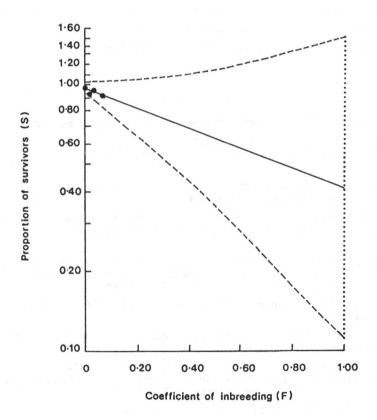

Figure 4:   Survival rate with inbreeding. (From Bittles & Makov, 1985.)
[Reproduced by kind permission of Taylor & Francis, Ltd.]

CONCLUSIONS

Besides the theoretical significance of human inbreeding studies, an accurate assessment of the risks associated with consanguinity has important practical consequences, for example, in genetic counselling and in legislation strongly influenced by genetic risks to the progeny. Therefore the enormous divergence in published estimates of excess mortality and/or morbidity in consanguineous offspring is a source of major concern. From the data presented, it can be seen that many of the earlier studies produced results which appear to have been considerably exaggerated. In part, this was due to the inadequacy of control groups used and to somewhat arbitrary modelling. The failure to apply appropriate statistical techniques has led to additional, inter-pretational problems. Thus the overall conclusion must be that, with the exception of incest and families known to carry deleterious recessive mutants, the risks to the offspring of inbred unions generally are within the limits of acceptability. For first cousin progeny, it also must be admitted that they appear to be in remarkable close agreement with the levels calculated by Darwin in 1875.

REFERENCES

Adams, M.S. & Neel, J.V. (1967). Children of incest. Pediatrics, **40**, 55-62.

Adinolfi, M. (1986). Recurrent habitual abortion, HLA sharing and deliberate immunization with partner's cells: a controversial topic. Human Reproduction, **1**, 45-48.

Agrawal, N., Sinha, S.N. & Jensen, A.R. (1984). Effects of inbreeding on Raven Matrices. Behavior Genetics, **14**, 579-585.

Alfi, O.S., Chang, R. & Azen, S.P. (1980). Evidence for genetic control of nondisjunction in man. American Journal of Human Genetics, **32**, 477-483.

Asha Bai, P.V., John, T.J. & Subramaniam, V.R. (1981). Reproductive wastage and developmental disorders in relation to consanguinity in South India. Tropical and Geographical Medicine, **33**, 275-280.

Baird, P.A. & McGillivray, B. (1982). Children of incest. Journal of Pediatrics, **101**, 854-858.

Bashi, J. (1977). Effects of inbreeding on cognitive performance. Nature, **266**, 440-442.

Bigozzi, U., Conti, C., Guazzelli, R., Montali, E. & Salti, F. (1971). Morbilita e mortalita nella prole di 300 coppie di coniugi consanguinei nel Commune di Firenze. Acta Genetica (Basel), **19**, 515-528.

Bittles, A.H. (1979). Incest re-assessed. Nature, **280**, 107.

Bittles, A.H. (1980). Inbreeding in human populations. Journal of Scientific and Industrial Research, **39**, 768-777.

Bittles, A.H. & Makov, E. (1985). Linear regressions in the calculation of lethal gene equivalents in man. Annals of Human Biology, **12**, 287-289.

Bittles, A.H., Radha Rama Devi, A., Savithri, H.S., Rajeshwari Sridhar & Appaji Rao, N. (1985). Inbreeding and post-natal mortality in South India: effects on the gene pool. Journal of Genetics, **64**, 135-142.

Böök, J.A. (1957). Genetical investigations in a North Swedish population. The offspring of first-cousin marriages. Annals of Human Genetics, **21**, 191-223.

Carter, C.O. (1967). Risk to offspring of incest. Lancet, **1**, 436.

Cavalli-Sforza, L.L. & Bodmer, W.F. (1971). The Genetics of Human Populations, p. 360. San Francisco: Freeman.

Darwin, G. (1875a). Marriages between first cousins in England and their effects. Journal of the Statistical Society, **38**, 153-184.

Darwin, G. (1875b). Note on the marriages of first cousins. Journal of the Statistical Society, **38**, 344-348.

Edo, M.A., Otero, H.R. & Caro, L. (1985). The influence of consanguinity on fertility and infant mortality in Sanabria (Zamora, Spain). Biology and Society, **2**, 129-134.

Ericksen, J.A., Ericksen, E.P., Hostetler, J.A. & Huntington, G.E. (1979). Fertility patterns and trends among the Old Order Amish. Population Studies, **33**, 255-276.

Fraser, F.C. & Biddle, C.J. (1976). Estimating the risks for offspring of first cousin matings. American Journal of Human Genetics, **28**, 522-526.

Freire-Maia, N. (1984). Effects of consanguineous marriages on morbidity and precocious mortality: genetic counselling. American Journal of Medical Genetics, **18**, 401-406.

Freire-Maia, N. & Elisbão, T. (1984). Inbreeding effect on morbidity. III. A review of the world literature. American Journal of Medical Genetics, **18**, 391-400.

Garrod, A.E. (1902). The incidence of alkaptonuria: a study in chemical individuality. Lancet, **2**, 1616-1620.

James, W.H. (1975). The fitness of the human zygote. Journal of Biosocial Science, **7**, 1-4.

Lindelius, R. (1980). Effects of parental consanguinity on mortality and reproductive function. Human Heredity, **30**, 185-191.

MacCluer, J.W. (1980). Inbreeding and human fetal death. In: I. H. Porter & E. B. Hook (eds.), Human Embryonic and Fetal Death, p. 241. New York: Academic Press.

Macfarlane, A. & Mugford, M. (1984). Birth Counts: Statistics of Pregnancy and Childbirth. London: Her Majesty's Stationery Office.

McKusick, V.A. (1978). Mendelian Inheritance in Man, 5th edn., p. xii. Baltimore/London: Johns Hopkins University Press.

Makov, E. & Bittles, A.H. (1986). On the choice of mathematical models for the estimation of lethal gene equivalents in man. Heredity, **57**, 377-380.

May, R.M. (1979). When to be incestuous. Nature, **279**, 192-194.

Medical Research Council (1978). Review of Clinical Genetics. London: Medical Research Council.

Miller, J.F., Williamson, E., Glue, J., Gordon, Y.B., Grudzinskas, J.G. & Sykes, A. (1980). Fetal loss after implantation. Lancet, **2**, 554-556.

Morton, N.E. (1960).   The mutational load due to detrimental genes in man.
    American Journal of Human Genetics, **12**, 348-364.

Morton, N.E., Crow, J.F. & Muller, H.J. (1956).   An estimate of the mutational
    damage in man from data on consanguineous marriages.   Proceedings of
    the National Academy of Sciences (U.S.A.), **42**, 855-863.

Mowbray, J.F., Gibbings, C., Liddel, H., Reginald, P.W., Underwood, J.L. &
    Beard, R.W. (1985).   Controlled trial of treatment of recurrent spontane-
    ous abortion by immunisation with paternal cells.   Lancet, **1**, 941-943.

Neel, J.V. (1958).   A study of major congenital defects in Japanese infants.
    American Journal of Human Genetics, **10**, 398-445.

Neel, J.V. & Schull, W.J. (1962).   The effect of inbreeding on mortality and
    morbidity in two Japanese cities.   Proceedings of the National Academy
    of Sciences (U.S.A.), **48**, 573-582.

Neel, J.V., Schull, W.J., Kimura, T., Tanigawa, Y., Yamamoto, M. & Nakajima,
    A. (1970a).   The effects of parental consanguinity and inbreeding in
    Hirado, Japan. III. Vision and hearing. Human Heredity, **20**, 129-155.

Neel, J.V., Schull, W.J., Yamamoto, M., Uchida, S., Yanase, T. & Fujiki, N.
    (1970b).   The effects of parental consanguinity and inbreeding in Hirado,
    Japan.   II. Physical development, tapping rate, blood pressure, intelli-
    gence quotient and school performance.   American Journal of Human
    Genetics, **22**, 263-286.

Rao, P.S.S. & Inbaraj, S.G. (1977).   Inbreeding effects on human reproduction in
    Tamil Nadu of South India.   Annals of Human Genetics, **41**, 87-98.

Rao, P.S.S. & Inbaraj, S.G. (1979).   Trends in human reproductive wastage in
    relation to long-term practice of inbreeding.   Annals of Human Genetics,
    42, 401-413.

Rao, P.S.S. & Inbaraj, S.G. (1980).   Inbreeding effects on fetal growth and
    development.   Journal of Medical Genetics, **17**, 27-33.

Record, R.G. & Armstrong, E. (1975).   The influence of the birth of a malformed
    child on the mother's further reproduction.   British Journal of Preventive
    and Social Medicine, **29**, 267-273.

Roberts, C.J. & Lowe, C.R. (1975).   Where have all the conceptions gone?
    Lancet, **1**, 498-499.

Romeo, G., Menozzi, P., Ferlini, A. et al (1983).   Incidence of classic PKU in
    Italy estimated from consanguineous marriages and from neonatal screen-
    ing.   Clinical Genetics, **24**, 339-345.

Romeo, G., Menozzi, P., Mastella, G. et al (1981).   Studio genetico ed
    epidemiologico della fibrosi cistica in Italia.   Rivista Italiana Pediatrica,
    7, 201-209.

Schull, W.J. (1958).   Empirical risks in consanguineous marriages: sex ratio,
    malformation and viability.   American Journal of Human Genetics, **10**,
    294-343.

Schull, W.J., Furusho, T., Yamamoto, M., Nagano, H. & Komatsu, I. (1970).   The
    effect of parental consanguinity and inbreeding in Hirado, Japan. IV. Fer-
    tility and reproductive compensation.   Humangenetik, **9**, 294-315.

Schull, W.J., Komatsu, I., Nagano, H. & Yamamoto, M. (1968).   Hirado:
    temporal trends in inbreeding and fertility.   Proceedings of the National
    Academy of Sciences (U.S.A.), **59**, 671-679.

Schull, W.J., Nagano, H., Yamamoto, M. & Komatsu, I. (1970). The effects of parental consanguinity and inbreeding in Hirado, Japan. I. Stillbirths and pre-productive mortality. American Journal of Human Genetics, **22**, 239-262.

Schull, W.J. & Neel, J.V. (1965). The Effects of Inbreeding in Japanese Children. New York: Harper & Row.

Schull, W.J. & Neel, J.V. (1972). The effects of parental consanguinity and inbreeding in Hirado, Japan. V. Summary and interpretation. American Journal of Human Genetics, **24**, 425-453.

Schull, W.J., Yanase, T. & Nemoto, H. (1962). Kuroshima: the impact of religion on an island's genetic heritage. Human Biology, **34**,271-298.

Seemanová, E. (1971). A study of children of incestuous matings. Human Heredity, **21**, 108-128.

Shami, S.A. & Zahida, (1982). Study of consanguineous marriages in the population of Lahore, Punjab, Pakistan. Biologia, **28**, 1-15.

Slatis, H.M. & Hoene, R.E. (1961). The effect of consanguinity on the distribution of continuously variable characteristics. American Journal of Human Genetics, **13**, 28-31.

Smith, C.A.B. (1967). Note on a paper by S. Kumar, R. A. Pai and M. S. Swaminathan. Annals of Human Genetics, **31**, 146-147.

Smith, C.A.B. (1969). Corrigenda. Annals of Human Genetics, **32**, 419.

Stern, C. & Charles, D.R. (1945). The Rhesus gene and the effect of consanguinity. Science, **101**, 305-307.

Stevenson, A.C., Davison, B.C.C., Say, B., Ustuoplu, S., Liya, D., Abul-Einen, M. & Toppozada, H.K. (1971). Contribution of fetal/maternal incompatibility to aetiology of pre-eclamptic toxaemia. Lancet, **2**, 1286-1289.

Stevenson, A.C., Johnston, H.A., Stewart, M.I.P. & Golding, D.R. (1966). Congenital malformations: a report of a study of series of consecutive births in 24 centres. Bulletin of the World Health Organisation, **34** (Suppl.), 88-102.

Stevenson, A.C. & Warnock, H.A. (1959). Observations on the results of pregnancies in women resident in Belfast. 1. Data relating to all pregnancies ending in 1957. Annals of Human Genetics, **23**, 382-391.

Sutter, J. & Tabah, L. (1952). Structure de la mortalité dans les familles consanguines. Population, **8**, 511-526.

Sutter, J. & Tabah, L. (1953). Fréquence et nature des anomalies dans les familles consanguines. Population, **9**, 425-450.

Whittaker, P.G., Taylor, A. & Lind, T. (1983). Unsuspected pregnancy loss in healthy women. Lancet, **1**, 1126-1127.

Yamaguchi, M., Yanase, T., Nagano, H. & Nakamoto, N. (1970). Effects of inbreeding on mortality in Fukuoka population. American Journal of Human Genetics, **22**, 145-159.

Yokoyama, S., Reich, T. & Morgan, K. (1981). Inbreeding and the genetic control of nondisjunction. Human Genetics, **59**, 125-128.

MATING PATTERNS AND GENETIC DISEASE

P. HARPER[1] and D. F. ROBERTS[2]

[1] Department of Medical Genetics, Welsh National School of Medicine, Cardiff, U.K.

[2] Department of Human Genetics, University of Newcastle upon Tyne, Newcastle upon Tyne, U.K.

INTRODUCTION

The object of this chapter is not to present fresh data or reanalyse what is in the literature, but to illustrate the clinical significance of mating patterns and how these may be relevant to the occurrence of genetic disorders. Medical genetic information is now very extensive and provides a worthwhile source for population biologists to work on. But there is the question of how to translate some of the ample data on consanguinity and abnormalities into the practical figures for risks which are needed in genetic counselling. Collaboration in this area would thus be mutually profitable, advantageous both to the human population biologist and the clinical geneticist.

HAEMOPHILIA

When Lady Bracknell told Gwendoline "An engagement ..... is hardly a matter that (a young girl) could be allowed to arrange for herself" she was merely echoing the belief held in innumerable societies throughout the world, and not only those of society in England at the turn of the century. This was particularly true of the Royal houses and the aristocracy with all the complex levels that were distinguished within it. Princess May of Teck, first cousin once removed of Queen Victoria, was regarded as almost unmarriageable, almost an outcast, to even the smallest reigning houses of Europe, not because of her parents' insolvency nor her own shyness, but because of the morganatic marriage of her grandfather with Claudine, Countess Rhedey, a Hungarian of ancient lineage but without the immediate royal ancestry that would have made her formally acceptable to a 19th century reigning house. Princess May was rescued from her plight by Queen Victoria, who did not share the antipathy of her continental cousins to a "flawed" pedigree, and who determined that here was an admirable wife for her grandson Prince Eddie and then, on his untimely premarital death, to Prince George, later to become George V. Concerned to make good marriages for her children that would strengthen both family and political ties, Queen Victoria arranged that four of her five daughters married into the royal houses of Europe.   And so she was responsible for distributing throughout Europe that unfortunate mutation that seems most likely to have occurred in one of her parents' gametes, hemophilia.   It so happened that it did not manifest until Queen Victoria's eighth child and fourth son, Leopold Duke of Albany, and then, via her first, second and fifth daughters, in ten of her grandchildren and great-grandchildren (Fig. 1).   Here then is one example of the way in which the mating practices of less than a century ago disseminated a deleterious gene throughout a particular stratum of society and throughout much of Europe, from Spain to Russia.

Here the dissemination of the gene was entirely due to the marriage pattern - these marriages and the subsequent movement of the participants would have occurred irrespective of the genes that they were carrying.   But sometimes it is the genes themselves that contribute to the occurrence of a marriage.   Mrs. J.L. inherited the hemophilia gene from her father, she married and produced an affected son.   Her marriage did not survive the ensuing stress, which so frequently occurs when a child with a serious genetic defect is born, and, wanting a normal son but not understanding that it was she who had transmitted the gene, she married again and had two more affected sons by her second husband.

HUNTINGTON'S CHOREA

Huntington's chorea is a serious neurological disorder (Hayden, 1981). Usually beginning in adult life, the patient develops involuntary movements, may become aggressive, there is steady deterioration of motor control, usually accompanied by general mental deterioration.   In some cases it is accompanied by increased libido and diminished self control, and illegitimate offspring of young adult onset patients sometimes make pedigrees harder to trace.   Only in a minority of cases is there any significant impairment of fertility or genetic fitness - those cases in which the onset is very early - and in fact most studies show the contrary.   It is one of those few disorders where there is an actual increase in genetic fitness regardless of whether one uses a control population socially matched, or whether one uses a comparison with unaffected sibs within the families, to demonstrate this.   It is unfortunately true that a proportion of marriages cannot stand the strain that the insidious, slowly progressive deterioration imposes, and the social and domestic deterioration that accompanies it. Sometimes it happens that a man who is affected may leave home and become a vagrant, and this has occurred in no fewer than 4 out of 100 Newcastle families,

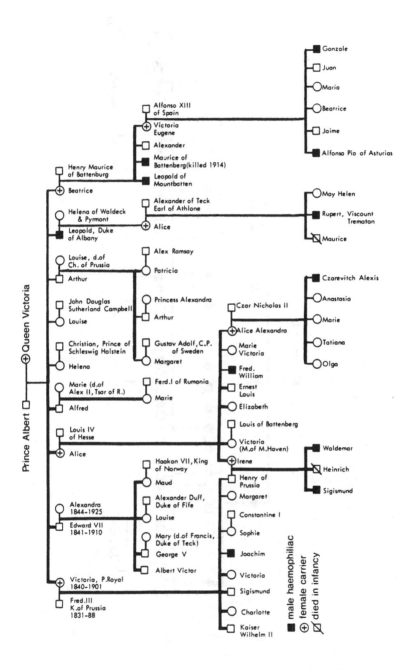

Figure 1: Hemophilia in Queen Victoria's family.

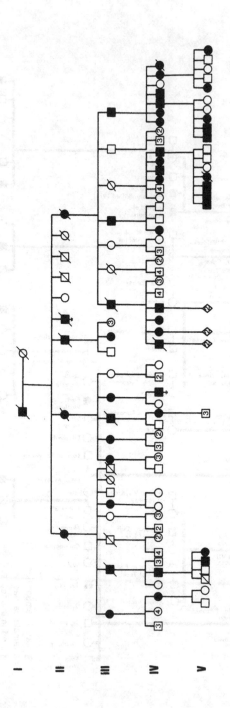

Figure 2: Pedigree of Huntington's chorea.

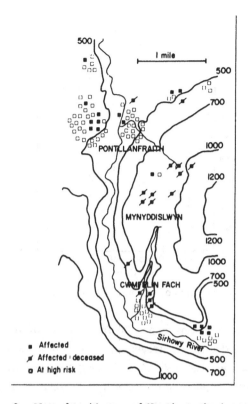

Figure 3: Map of residences of Huntington's chorea family.

where it was the man who was affected and where the onset was in the early or middle 30's, and the wife was left with the children in very difficult circumstances.

In south Wales there is indeed a high prevalence of Huntington's chorea (Spillane & Phillips, 1937). The current prevalence in Gwent is 9.52 per hundred thousand, and undoubtedly the gene has been common and widespread in south Wales for many years. But Huntington's chorea is not evenly distributed throughout south Wales—there is a notable focus of the disease in the two most easterly mining valleys (the Sirhowy and Afon Llwyd). Most cases in each valley belong to a single kindred. The kindred illustrated in Figure 2 derives from a single immigrant, who was born in Devon in 1854 and came to live in the hamlet of Mynydd Islwyn in the farmhouse on the ridge above the Sirhowy valley. He was by trade a mason, being attracted to south Wales by the rapid development that came with the opening of the coal mines. His descendants were numerous, they remained in the valley and most of them live in the immediate area (Figure 3) - with the exception of about 4 individuals, all those in that first

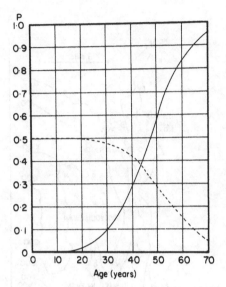

Figure 4:   Distribution of age at onset of Huntington's chorea

pedigree can be located on this map.   So far 46 cases of Huntington's chorea have resulted from this single progenitor, and today there are 110 living descendants who are at high risk of developing the disorder.   This family illustrates what seems to occur in the majority of cases.   Huntington's chorea is a highly immobilising disease for the family.   The possession not just of the disease, but perhaps of a parent who has the disease, seems in this case to have prevented at least in part the movement outwards from this very restricted district.   Whereas the general population as a whole has become highly depleted and there has been considerable emigration as the industries have declined, all these individuals remain clustered tightly around this area.

To the high prevalence of this disorder, this kindred and a few others have contributed disproportionately.   The 132 living cases of the disease belong to 98 separate kindreds, although many of these might well prove to have common ancestry if their pedigrees were thoroughly traced.   The pedigree (Fig. 2) shows the characteristic dominant pattern of inheritance.   In south Wales the number of cases proven to be due to new mutations is zero.   This does not mean in fact that none of the cases was due to a new mutation, but just that proving a new mutation in a late onset disorder is extremely difficult, for the parents have to be available to be examined, not having died from other causes.   Figure 4 shows the distribution of actual age of onset, the continuous curve giving the probability that somebody actually possessing the gene will have shown the disease by a particular age; the dotted line shows the corresponding probability that somebody with an affected parent is actually carrying the gene (Harper et al, 1979). This age distribution provides a major problem in genetic counselling of families, because at the ages where the information is required, when such an enquirer is about to start family building, the risk remains very close to the 50% that was applicable when he was born, though this situation is beginning to change as a result of recent developments in presymptomatic diagnosis. It also inhibits geographical mapping of the gene and affects potential marriages of family members - in established communities young people may well be cautioned against marriage with a member of "that" family.

This example shows how a deleterious gene may affect movement in the population; it may prevent marriage or lead to its breakdown; nevertheless it tends to enhance rather than reduce fertility.

PHENYLKETONURIA

The next example derives from a chance discovery during the screening programme in Wales aimed primarily at detecting phenylketonuria. From every child born a small sample of blood is obtained from a heel prick on to a piece of filter paper, which is examined for the level of a particular amino acid, phenylalanine, and this screening test is a valuable indicator of the possible presence of the disease phenylketonuria (Guthrie & Susi, 1963). With every such sample that arrives at the laboratory is included fundamental data such as exact place of residence, date of birth, sex, so that this material provides a very valuable source for population studies, without the bias that one may find for example in a sample of blood donors. Figure 5 shows the pedigree of one affected baby who was detected. It depicts a complex mating pattern, and relates to a gypsy family. Following this finding, Dr. Mair Williams worked with the gypsies for a period of time and obtained an enormous amount of pedigree data as well as blood samples (Williams & Harper, 1977; Harper & Williams, 1975). The Welsh gypsies proved to be highly inbred (Williams, 1986). This pedigree is but one of several. Moreover the parents of that first child were not aware how closely related they were - they just knew that basically they were both gypsies and that was all that mattered. It was indeed a suprise to find this highly inbred population at our door.

Phenylketonuria was shown to be at high prevalence, not only in this kindred but throughout the Welsh gypsies, 1 in 240 by contrast to the average of 1 in 10,000 in England; there is a gradient of diminishing frequency south eastwards across the United Kingdom, with an incidence in London of about 1 in 25,000 (Carter, 1973). This is not a Welsh disease, for the incidence in Wales is

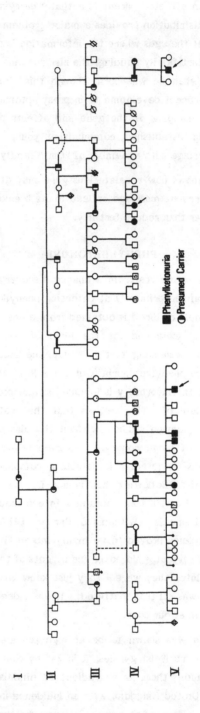

I

II

III

IV

V

Figure 5: Pedigree of gypsy baby with phenylketonuria.

closely comparable to that in adjacent areas in the Midlands. There is no predominance of Welsh surnames in the patients, nor do their grandparents show any excess of Welsh-born individuals. The frequency is higher in Scotland and Ireland, but in Wales it seems to be distinctive of the Welsh gypsies (Harper, 1986).

Gypsies in Britain are quite numerous (15,000) but this figure includes Irish tinkers and other wandering groups. In Wales there are thought to be about 1500 gypsies, most of them in the south. Several gypsy kindreds in Wales have been extensively documented. For example, there was the Wood kindred of north Wales (Sampson, 1926) which included many gifted members. They were trilingual in Welsh, English and Romany, were great musicians, who performed for Queen Victoria in Beaumaris, and several of their descendants have attained distinction (e.g. Augustus John). But so far there has not been any comprehensive analysis of the gypsy pedigrees to assess the extent of inbreeding, or how far it contributes to the manifestation of phenylketonuria and other recessive disorders. Certainly the endogamy that is characteristic of the Welsh gypsies and the inbreeding to which it gives rise, is to some extent responsible for the high incidence of phenylketonuria, and other recessive disorders. It has long been known that of all patients with recessive disorders, the proportion born to the offspring of first cousin unions varies with the gene frequency and with the frequency of first cousin unions as in the following equation:

$$k \quad = \quad \frac{c(1 + 15q)}{16q + c(1 - q)}$$

where k is the proportion of recessives born to first cousin unions, c is the incidence of first cousin marriage in the population, and q is the gene frequency (Dahlberg, 1948).

Here then is the classic example where the mating pattern of the population, namely the practice of inbreeding, leads to the excessive manifestation of a rare recessive disorder, where a pair of homologous deleterious genes is required at the relevant locus. Inbreeding of course has no effect on incidence of dominant traits. Since in these only a single gene is required to produce a dominant trait, the bringing together of identical alleles in the offspring of consanguineous unions will not cause the trait to manifest a greater frequency.

OTHER DISORDERS

Unlike the disorders discussed so far, most congenital malformations are not attributable to single gene inheritance, nor are many other developmental disorders and pathological conditions. Chromosomal conditions are associated with the mating patterns essentially in so far as the latter affect parental ages.

Populations characterised by late marriage or with major variations in the distribution of births by maternal age will be expected to show variations in the incidence of Down's syndrome, for the risk of this and other disorders caused by non-disjunction increases with advancing maternal age. An example in Britain is the incidence of Down's syndrome in the Shetland Islands, the highest in Britain, which is almost entirely attributable to the number of women who continue to have babies into their forties, i.e. the older mean maternal age of the population.

There are many disorders of multifactorial aetiology, where the genetic contribution is polygenic. Their aetiology is understandable on the threshold model proposed by Falconer (1965). There is an attribute in the population, assumed to be normally distributed, of liability to develop a particular condition. Those who manifest it fall beyond a threshold in the distribution. The genetic component in this liability is polygenic. Polygenes are transmitted in the same way as, and in accordance with the same laws as, major genes, but their effects do not provide sufficient discontinuity for individual study. A polygene acts as one of a system, the members of which may act together to effect large phenotypic differences, or against each other so that similar phenotypes develop from different genotypes. An individual polygene has only a slight effect, is apparently interchangeable with others within the system, does not have an unconditional advantage over its allele, since its effect depends on other alleles present in the system, and it cannot therefore be heavily selected for or against. Such a system conceals great genetic diversity. The result in the individual depends more on the loading of genes increasing liability that he carries than on the particular genes present. Then the whole system is capable of modification by the environment, which may shift the curve of liability along the X axis or alter its dispersion and so cause a greater or smaller proportion of the population to fall beyond the threshold.

Of the limited number of conditions so far studied on the south Wales gypsy population, non-specific mental subnormality is one candidate for this inheritance model. Out of a total of 129 live births produced by 43 consanguineous unions, there were 19 cases of non-specific subnormality. Out of 149 live births from 56 non-consanguineous unions, there was only 1 case, a highly significant difference ($\chi^2 = 19.0$; $p < .001$). Another example relates to spontaneous abortions and stillbirths. There were 11 of these reported from the consanguineous unions as compared with three from the non-consanguineous, again a significant excess ($\chi^2 = 5.8$; $p < .02$).

In a population where the practice of inbreeding is the characteristic mating pattern, one would expect to see an increased incidence of such multifactorial disorders. The mechanism would be the same as for Mendelian recessives, namely the bringing together of identical alleles in the offspring of consanguineous unions, so that there is identical homozygosity at a greater number of the polygenic loci than in non-inbred individuals.

PROBLEMS

These examples illustrate some of the interrelationships existing between genetic disorders and mating practices. The real problems come in applying the deductions from them. There are many disorders where the inheritance is not known. There are of course many disorders well established as recessive from very extensive and very clear family data. But it remains true that the diagnosis of a recessive disease in a counselling situation is rarely made on the basis of the family history of the patient but usually from the details of its clinical manifestation. Moreover, many syndromes have only recently been recognised and in Britain the incidence of some malformations may represent a fairly recent state. Many malformations were lethal until very recently, and many still are. But there have been very big changes in pediatric practice, so that the idea that a child with a severe malformation should be allowed to die is now a thing of the past, and most neonatal units will actively resuscitate a child with a defect who would otherwise die. Whether one agrees or disagrees with this, within Western societies it has resulted in an increase in the number of affected children who survive long enough to allow diagnosis to be made.

In a counselling situation one may be faced with a couple from an inbred population, such as the gypsies; it may be a marriage between two members of the population at large, or between first cousins. Their child has an abnormality which is not recessively inherited, or which one finds difficult to identify. To what extent is the occurrence of the disorder related to consanguinity of the parents? These are two different situations. In the first, an inbred population may contain many spouse pairs whose kinship is quite remote, and indeed if there has been only quite occasional admixture into the population, the inbreeding coefficient of their offspring may be negligible and so lead to little increased risk. In the second, the offspring will be identically homozygous at at least one in every sixteen loci. So one must beware of attributing any disorder seen in a member of an inbred community to the inbreeding that has occurred.

How far does the excessive risk associated with Mendelian recessive disorders in an inbred community extend to disorders with other modes of inheritance? Here the difficulty comes with the multifactorial conditions in which there is environmental lability. Certainly the excess mortality and morbidity and multitude of abnormalities suggested by earlier studies of inbred communities cannot be totally attributed either to single genes of recessive effect or pairs of polygenes, for there are many other relevant factors.   One often finds that the conditions that lead to choice of a related spouse are also associated with degree of intelligence, socioeconomic status, diet, religion, restricted population size, which may affect the liability to develop a disorder.

In dealing with an inbred population, does consanguineous marriage within the population actually increase or decrease the risks to offspring?   There has been much discussion whether a population that is closely inbred for a long period has bred out the deleterious recessives over the generations.   This argument has been particularly strong in relation to the south Indian populations where uncle-niece marriages are common.   Unfortunately some of the surveys that purport to show a reduced incidence of recessive disorders can be criticised on the grounds of method, e.g the surveys may have been made by unqualified personnel unable to identify recessive syndromes.   Certainly when attempts have been made to investigate objectively the prevalence of disorders by say urine analysis of total populations of new born babies, the aminoacidurias appear in the expected frequencies (Bittles et al, 1986).

If a patient is seen to have a disorder in an inbred population, to what extent should it be attributed to the population consanguinity?   There is a strong tendency to attribute any disorder seen in a member of an inbred community to the inbreeding that has occurred.   But many inbred communities are small, have existed in isolation from other populations perhaps for a prolonged period, and other possible explanations for the occurrence of high frequencies of some disorders include enhanced drift that comes with small population size, the founder principle and subsequent expansion from a small group, possibly selection.   One needs to eliminate such variables before attributing the high frequency of disorders to consanguinity or inbreeding in a population as a whole.

CONCLUSION

In Britain today there are a number of groups who are inbred, not only the long established peoples such as the gypsies and some of the island populations, but the more recently immigrant ethnic groups who practise cousin marriage. Some of these, as a result of their relative isolation in this country from other family members, and hence more restricted choice of spouse, appear to be becoming more inbred then they were in their native land.   In some localities ethnic minority families are seen quite frequently in genetic counselling clinics, where they present with many unusual or unidentified disorders, some of which are certainly recessive. So the type of problem outlined here is more than academic.

There is a lot of work to be done in terms of translating general surveys into specific figures that can be applied. The medical data are interesting, fascinating, if often rather imprecise. It is hoped that this brief discussion of some features of the interrelationship between genetic disorders and mating practices may show how rewarding detailed collaborative effort would be.

REFERENCES

Bittles, A.H., Radha Rama Devi, A. & Appaji Rao, N. (1986) Inbreeding and the incidence of recessive disorders in the populations of Karnataka, South India, In: D. F. Roberts & G. F. De Stefano (eds.), Genetic Variation and its Maintenance, pp. 221-228. Cambridge: Cambridge University Press.

Carter, C.O. (1973) Nature and distribution of genetic abnormalities. J.Biosoc.Sci. 5, 261-272.

Dahlberg, G. (1948a) Mathematical Methods for Population Genetics. New York: Interscience Publishers.

Falconer, D.S. (1965) The inheritance of liability to certain diseases, estimated from the incidence among relatives. Ann.Hum.Genet., Lond. 29, 51-76.

Guthrie, R. & Susi, A. (1963) A simple phenylalanine method for detecting phenylketonuria in large populations of newborn infants. Pediatrics, 32: 338.

Harper, P.S. (1986) Mendelian disorders in Wales, In: P. S. Harper & E. Sunderland (eds.) Genetic and Population Studies in Wales, pp. 273-289. Cardiff: University of Wales Press.

Harper, P.S., Walker, D.A. & Tyler, A. (1979). Huntington's chorea. Lancet, ii, 346-349.

Harper, P.S. & Williams, E.M. (1975) Phenylketonuria in gypsies. Lancet, i, 1041.

Hayden, M.R. (1981) Huntington's Chorea. Berlin: Springer Verlag.

Sampson, J. (1926) The Dialects of the Gypsies of Wales: Being the Older form of British Romani preserved in the Speech of the Clan of Abram Wood. London.

Spillane, J.D. & Phillips, R. (1937) Huntington's chorea in South Wales. Quart.J.Med. 6, 403-423.

Williams, E.M. (1986) Genetic studies of Welsh Gypsies, In: P. S. Harper & E. Sunderland (eds.), Genetic and Population Studies in Wales, pp. 186-211. Cardiff: University of Wales Press.

Williams, E.M. & Harper, P.S. (1977) Genetic study of Welsh gypsies. J.Med.Gen. 14, 172-176.

PART IV

SOCIAL, RELIGIOUS AND CULTURAL FACTORS

THE EFFECT OF PREFERENCE RULES ON MARRIAGE PATTERNS

B. DYKE[1] and P. G. RIVIÈRE[2]

[1] *Department of Genetics, Southwest Foundation for Biomedical Research, San Antonio, Texas;*

[2] *Institute of Social Anthropology, University of Oxford, Oxford, U.K.*

INTRODUCTION

Complex symbolic and cognitive abilities are assumed to be the major factors leading to the great elaboration of patterned behaviours that appears to set humans apart from other organisms.   Nowhere is this patterning more evident and more potentially important than in human mating behaviour.   The purpose of this contribution is to illustrate an approach to the analysis of mating patterns that differs considerably from the usual analysis of phenotypic assortative mate choice in that its purpose is to evaluate the influence of marriage rules and preferences on observed marriage patterns in a single small population.

The rationale underlying this method rests first on the observation that demographic factors such as population size, age-sex structure and distribution of family sizes will determine for any individual the number of potential mates. Further, we assume that marriages between any class of potential mates may arise either by chance (that is, by random mating) or as the result of deliberate choice (that is, by preferential mating).   Finally we assume that holding demographic factors constant, a population adhering strictly to a scheme of random mate choice would exhibit marriage patterns different from one in which strong preferential mating was the practice.

The test of marriage preference consists of comparing observed proportions of marriages classified by age difference, kinship and residence of the couple against expectations derived from computer simulations of two possible schemes of mate choice.   The first scheme assumes that marriage is made at random with respect to age, genealogy and residence.   Preferred marriages that result from mate choice of this sort will depend simply on opportunities defined by the demographic structure of the population.   The second scheme assumes that marriage with preferred classes results from deliberate choice of an appropriate partner from the pool of potential mates.

The data used in this study come from the Trio, a group of horticultur-
ists living in an area at present divided between Brazil and Surinam.
Genealogies and vital statistics as well as information on marriage preferences
were collected from these people during the course of fieldwork in 1963 (Rivière,
1969) and again in 1978. Missing birth and death dates were estimated for some
individuals using the method described in Lynch et al (1983) based on a recent
demographic analysis (Gage et al, 1984a, 1984b). The work described here
relates to two Surinam Trio villages totalling about 400 individuals as they
existed in about 1963. In the decade ending in that year, 258 marriages were
recorded between 237 men and 228 women. This work is an extension of an
earlier study (Fredlund & Dyke, 1976) using improved methods and more detailed
and extensive genealogical data.

The Trio are explicit in expressing preferences for and prohibitions
against marriage between individuals of certain genealogical relationships as
shown in Table 1. Endogamy is also an explicitly stated preference. Although
not explicitly stated by the Trio, analysis of the marriages revealed that the
distribution of age differences between spouses was not random, as can be seen
from Figure 1. We have treated age difference as a preferential rule.

Table 1: Trio preferred and prohibited kin categories (male perspective)

Preferred relationships*				
ZD	FBDD	MZDD	FMBSD	MFZDD
FZD	MBD	FFZSD	FMZDD	MMZSD
FZSD	MBSD	FFBDD	MFBSD	MMBDD

Nuclear family				
M	D	Z	FM	MM

Other prohibited relationships				
FFM	DDD	FZDD	FZDDD	MFZSD
FMM	BD	MZ	FMZ	MFBDD
MFM	BSD	MZD	FMBD	MMZ
MMM	ZSD	MZSD	FMZD	MMBD
SD	BDD	MBDD	FMZSD	MMSD
DD	ZDD	FFZ	FMBDD	MMBSD
SSD	FZ	FFBD	MFZ	MMZDD
DSD	FBD	FFZD	MFBD	
SDD	FBSD	FFBSD	MFZD	

* B = brother; D = daughter; F = father;
M = mother; S = son; Z = sister.

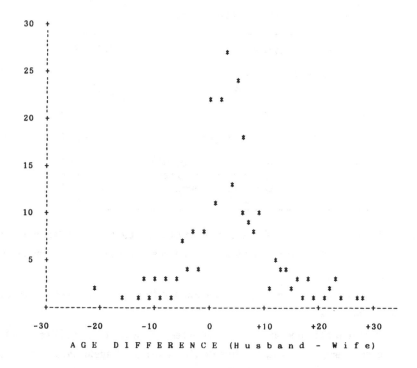

Figure 1: Distribution of age difference between spouses.

The simulation procedure repeatedly used samples at random from the same group of 237 men and 228 women to form couples according to various preferences and prohibitions imposed at the beginning of each replicate, using the following method:

1. The order of the male and female lists is randomised.

2. The first available individual is selected from the list of males. If he has already been "married" in the simulation process he is passed over and the next male is chosen.

3. The first available single female is selected and her birth year matched against that of the male. If her mating period does not overlap with that of the male or if the difference in ages of the couple is outside the range specified at the outset of the simulation, she is rejected and the next female is selected.

4. Kinship of the couple is determined by matching for the identity of the appropriate ancestors - for example, the couple are brother and sister if they share parents, first cousins if they share a set of grandparents, etc.

Table 2:   Simulation input parameters

Calendar years analysed	1954–1963
Marriage age for males	15 to 45
females	14 to 40
Number of replicates	50
Limits on age difference of spouses	Y/N
Marriage with prohibited kin permitted	Y/N
preferred kin required	Y/N
co-residents required	Y/N
non-preferred partners	Y/N

A female is rejected if she is a member of the male's nuclear family or of a more distant prohibited class of relatives specified in the simulation input.

5.   Joint residence of the couple is determined from natal residence codes of each individual.

6.   Rules of preference are applied to determine whether or not the couple are to "marry". If not, the female list is searched again until a mate is found who most closely satisfies the rules as specified by the simulation input.

7.   Various attributes of the marriage (such as difference in ages, kinship and residence of spouses) are recorded; both individuals are removed from their respective lists and the next available male is examined.

8.   When matching of the last male has occurred, marriages are tabulated and summarised.   Steps 1 to 7 may be repeated as many times as desired and the results of all replicates are averaged before the summary is produced.

Input to the simulation defines the population and specifies the conditions that govern the generation of artificial marriage histories. This can be seen in Table 2.

This analysis included all individuals who were of marriageable age in the decade 1954 to 1963. These ages (determined from the actual marriages of the period) were 15-45 for males and 14-40 for females. The remaining parameters made it possible to combine marriage rules in various ways so that the effects of each can be understood. For example, the simulation can be run so that the distribution of age differences between spouses will be the same as that shown for the actual marriages (that is, preferentially with respect to age) or it can be run without control of age differences (that is, randomly with respect to age). Likewise marriage with prohibited kin (excluding the nuclear family) can be permitted or not and each male can be forced to look through all available females for preferred kin or for co-residents or not as desired. It is also possible to permit or prohibit marriage between couples for whom no relationship can be found.

For many reasons a major interest in mating patterns lies in their potential effect on levels of consanguinity and migration. Because numbers of marriages are not large, consanguinity was measured very crudely by totalling the number of unions (both simulated and real) made with preferred kin and expressing this number as a proportion of all marriages made. The proportion of marriages made between co-residents was used as an estimate of migration.

The number of possible combinations of marriage rules that might affect these measures is quite large, making it quite important to simplify the analysis if possible. Fortunately for the Trio during this decade, the distribution of age differences of spouses had little effect on either the proportion of marriages made with kin or with co-residents, so it was possible to eliminate this parameter from further experiments. It was then possible to estimate the interaction of preference for kin and co-residence as shown in Table 3. Here, four sets of simulation replicates were made under differing sets of mating rules.

The first column in that Table shows proportions of marriages with preferred kin and with co-residents under random mating - that is, although marriages are tabulated according to kinship and residence the female list is not searched in its entirety for a mate who best satisfies the rules of preference (except for avoidance of prohibited kin and inappropriate age).

Table 3: Proportions of simulated marriages with preferred mates under selected rules of preference

	Preferences applied			
	None	Kinship Alone	Residence Alone	Kinship + Residence
Proportion of marriages with				
preferred kin	.003	.079	.007	.087
co-residents	.324	.372	.935	.981

Table 4:   Evaluation of Trio marriage rules governing preferred
               kin and endogamy

	Simulation			
	Preferred (P)	Random (R)	Observed (Q)	Index a
Kinship	.087	.003	.050	.559
Endogamy	.981	.324	.892	.864

$$Q = aP + (1.0 - a)R$$

The second column shows these proportions under a mating rule of partial preference which ignores residence but requires that a mate of preferred kin relationship be chosen whenever possible from the list of potential mates. The third column shows the figures when preference for endogamy alone is specified. The last column shows the effects of a fully preferential rule where we have specified that each male first searches the list of females for a mate of the appropriate kin type. If he is unable to find such a mate he tries to find a co-resident spouse; failing this, he will take the first unprohibited female of an appropriate age. Comparison of these results shows that the rule specified in each of the cases of partial preference forces the proportions quite closely toward the fully preferential figures. Conversely, where the rule is ignored proportions resemble those that hold under random mating. Clearly the rules of preference for kin are largely independent of those for residence for the Trio in this period. This justifies independent assessment of the strength of the rules as given in Table 4.

This table shows a comparison of observed proportions of actual Trio marriages classified by kinship and residence (the column headed by 'Q') against expectations derived from simulation of random and preferential mate choice (columns headed by 'R' and 'P' respectively). Clearly, in the case of both kinship and residence the proportion of marriages observed falls between the values resulting from the two simulated schemes of mate choice. The last column (headed by 'a') gives an Index of Mating Preference which is a simple weighting factor that estimates the extent to which observed marriages are the result of preferential or random mating.

Interpretation of this Index depends of course on the uses to which it is to be put. From a strictly behavioural perspective it can be thought of as the probability that a male will reject a non-preferred female as a mate, and can be used as a measure for making comparisons between populations or periods. In this study it can be seen that the Trio are rather severely constrained from marrying preferred kin. Given the age-sex structure of the population and known genealogical relationships only about 9% of all marriages could have been made between preferred relatives if mating preferences were followed as closely as possible. Despite these limited opportunities it is possible to distinguish the relative contribution of random and preferential mating with respect to kinship and in fact to show that preference plays an important role in understanding levels of consanguinity in the population. Likewise, preference for co-resident spouses plays an important part in establishing rates of endogamy.

Other questions of interest are related to changes in Trio society brought about by the increasing influence of Protestant missionaries. One issue is whether or not the Trio are adhering to their rules to the same extent in 1978 as they were in 1963 which we can tell simply by comparing the Index for the two periods. More detailed changes in the mating patterns can be determined by comparing relationships between different preferences such as those for kinship and residence (as shown here). We have also been investigating the effects of a preference for sib exchange which the Trio state to be next in order of preference if appropriate kin or co-resident mates cannot be found.

From a methodological perspective, we see benefits in quantifying mating rules for purposes of modelling demographic and genetic structure, particularly for studies of the accumulation of inbreeding. The Index of Mating Preference as it is defined can be used practically without modification to produce quite realistic mating patterns in a complex microsimulation. We also point out that, although the discussion here has been limited to somewhat simplified features of mating behaviour, the method can be conveniently extended to any pattern of assortative mating for which a rule can be defined and evidence for its application can be tabulated.

ACKNOWLEDGMENTS

This study was funded by U.S. National Institutes of Health Grant HD 17133. PGR wishes to acknowledge financial assistance from the Research Institute for the Study of Man in New York towards fieldwork in 1963-64, and from the Social Science Research Council of Great Britain towards fieldwork in 1978. Collaboration of the authors was also aided through funds provided by the Faculty of Anthropology and Geography, University of Oxford.

REFERENCES

Fredlund, E.V. & Dyke, B. (1976).   Measuring marriage preference.   Ethnology, 15, 35-45.

Gage, T.B., Dyke, B. & Rivière, P.G. (1984a).   Estimating mortality from two censuses: an application to the Trio of Surinam.   Human Biology, 56, 489-501.

Gage, T.B., Dyke, B. & Rivière, P.G. (1984b).   The population dynamics and fertility of the Trio of Surinam: an application of a two census method. Human Biology, 56, 691-701.

Lynch, C.M., Dyke, B. & Rivière, P.G. (1983).   Estimating vital rates for incomplete pedigrees.   Human Biology, 55, 63-72.

Rivière, P.G. (1969).   Marriage among the Trio.   Oxford: Clarendon Press.

RELIGIOUS RULES, MATING PATTERNS AND FERTILITY

V. REYNOLDS

Department of Biological Anthropology, University of Oxford, Oxford, U.K.

INTRODUCTION

Membership of a religious group, in general, implies acceptance of a set of shared beliefs and an associated set of rules about right and proper ways of thinking and behaving. This is very evident in the field of mate selection and marriage, and as I shall show has important implications for the understanding of fertility.

The major world religions are Judaism, Christianity, Islam, Hinduism and Buddhism. Their rules, first enunciated by their founders and prophets, have from time to time been codified and published in the great religious works, and these have been the source of further tracts and commentaries down the ages. For Islam, the Koran is wholly sacrosanct, the actual word of Allah, and commentaries tend to be conservative, or if radical to lead to a fundamentalist backlash. For Judaism and Christianity, modernisation has always been occurring, with occasional reversions to fundamentalism. Hinduism is and perhaps always has been very diverse, with a panoply of gods, and an enormously detailed set of rules relating to the caste system and the problem of pollution (Dumont, 1970). It is very traditional and resists the modern challenge of caste barriers emanating from Delhi and the Congress Party. Buddhism again takes various forms but perhaps because of its other-worldliness has survived alongside materialism and modernisation in Japan. Elsewhere, as in Burma, it maintains a traditional way of life with detailed rules of everyday behaviour.

Among the rules found in most, perhaps all religions, are those stressing endogamy within the religious community. Why should this be so? One undoubted reason is maintenance of the religion itself. Religious rules stress the need to maintain the human stock of their adherents. In their early years all the great religions had to marry out and convert in, and most still contain rules about this: Judaism, Catholicism and Islam allow out-marriage with in-conversion, although in the case of Islam it is only the man who is allowed to marry out. Such rules are characteristic of expansionist religions; Hinduism by contrast disapproves of out-marriage, especially with Islam, and conversion to Hinduism is difficult, although historically it has occurred, for example in Java and Bali.

Besides the existence of rules to ensure continuity of belief, there are other reasons why endogamy rules should be obeyed. One such reason is psychological and social: marriage is in many ways simpler if the partners share ideas in common and fit into a community of believers. There are also strong economic reasons for endogamy. Because marriage is often primarily concerned with property transmission and inheritance, families involved in marriage partnerships tend to be well matched in terms of status and wealth. If endogamy is close, it may serve to prevent fragmentation of landholdings. But in any case shared values help to ensure the continuance of the families' wealth. This is clearly seen in the case of some of the religious isolates of North America.

RELIGIOUS ISOLATES

The Hutterites

The Hutterites are an Anabaptist religious sect living in hamlets, or small colonies, in the USA and Canada. They are well-known for their extremely high rate of reproduction; a recent re-calculation gave a mean value of 8.85 children for the completed fertility of all women over age 45 in a sample of 340 women studied in 1950 (Lang & Goehlen, 1985).

This high fertility can be attributed to peculiarities of the mating system at the family level which in turn are based on Anabaptist ideas. The example they provide is of interest because it illustrates the role of religious ideas in determining fertility (Eaton & Mayer, 1953). To understand these ideas we need first to examine the history of the sect.

The Hutterites who entered the Dakota plains were members of a Protestant sect originating in Switzerland and Bohemia in 1528. Its leaders were prolific tract-writers during the Reformation. They insisted that infant baptism was contrary to Christian principle, and that only those baptised again as adults had truly entered the Christian faith. Persecuted at home, they were nearly exterminated by 1762 but survived in Russia and then went to the USA in the 1870s. In South Dakota they established hamlets organised on a communal basis, and continued to speak their original German. Few of them ever left their colonies in subsequent years and marriage was almost universally endogamous. All property was shared and a spiritual value was placed on austere living.

As Eaton and Mayer (op. cit.) state:

"In most cultural settings, a rapid population growth like that experienced by the Hutterites would tend to be checked by difficulties of providing the offspring of most families with the essentials of food, clothing, shelter and medical care. The Hutterite economy and value system are uniquely designed to deal with this problem. Even an inefficient or ill father can support a large family, while the more capable members of the group forgo the economic benefits which would accrue to them if they were to be paid wages for their labor based on their importance to the group effort. They do this out of religious conviction and ethnic loyalty. It is doubtful that the Hutterite reproductive record could be maintained were competing American materialistic and secular values to bring about any major modification in their communal economic structure." (p. 218).

Communal living and sharing of property and goods are still the Hutterites' practice today. For instance, food is consumed communally, agricultural work is communal, children are looked after communally, and profits from the sale of produce are held by a corporation for the welfare of the whole colony. The average Hutterite colony consists of four longhouses with four families in each. A typical family has a middle room with two adjoining bedrooms, and a storage space in the communal attic. The whole colony can in some ways be seen as an extended family, but the idiom of the family is religious: a spiritual brotherhood and sisterhood embracing all members of the community (Hostetler, 1974).

Despite the high rate of reproduction, age at marriage is not early. It is customary to marry only after adult baptism has taken place, and the median age of marriage for women in 1950 was 22 years, while for men it was 23.5 years. Peak fertility occurred in the 25-29 year age group, with an average birth rate of one birth every two years. Continued reproduction was culturally encouraged, few remained unmarried of either sex, divorce and desertion were rare. Intercourse was taboo before marriage, thus illegitimacy was very rare (10 known cases between 1875 and 1950 in a population numbering 8,542 in 1950). The only rules governing sex within marriage forbade intercourse shortly before and for six weeks after parturition, so that pregnancy was quickly resumed. The classlessness of Hutterite society, and their common standard of living, based on their doctrines, led to an absence of reproductive differentials between families. And finally, contraception was taboo, being referred to as "murder" and answerable for as such in the world to come; coitus interruptus was a sin, and the rhythm method was scarcely known.

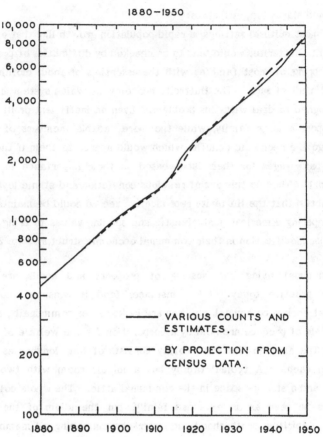

Figure 1:   Growth of the ethnic Hutterite population, 1880-1950. (From Eaton & Mayer, 1953.)

It was characteristics such as these, embedded in the group's religious ideology, that led to their great expansion (Figure 1).   Their faith was pronatalist, or what Reynolds and Tanner (1983) called "rc+".

Between the period 1946-50 and the year 1971 there was a decline in Hutterite fertility (see Table 1), though it is still very high (Laing, 1980).   Crude birth rate declined from 45.9 to 38.4, a drop of 16.3% in 21 years.   All age groups were affected, except the 15-19 years group in which the rate rose.   Mean age at marriage rose during this period quite markedly, from 22 to 24.9 years for women, and from 23.5 to 26 years for men.   The proportion of women aged over 30 years and still unmarried also rose sharply, from 54.% to 14.8%. Crude death rate and infant mortality rate declined.   Laing's estimates (based on standards for Alberta) show a fall of CDR from the 4.43 reported by Eaton and Mayer (op. cit.) to 3.85.   Infant mortality declined from 45 to 15.6.   Laing's conclusion

Table 1:   Hutterites: demographic changes, 1950-1971

	1950	1971
Crude birth rate (per 1000)	45.90	38.40
Total fertility rate	8.27	5.09
Crude death rate (per 1000)	4.43	3.85
Infant mortality rate (per 1000)	45.00	15.60
Mean age at marriage (M) (years)	23.50	26.00
Mean age at marriage (F) (years)	22.00	24.90
Remaining unmarried, over 30 years (F) (per cent)	0.54	1.48
Rate of natural increase (per 1000)	41.50	35.50

(From Eaton & Mayer, 1953, and Laing, 1980.)

is that the fall in CBR can be wholly accounted for by the increased age at marriage of women. This should in theory have led to a 29% fall in CBR. The actual fall was 25%. The difference can be accounted for by the unexpected increase in fertility of the 15-19 years age group, and possibly also by an increase in use of contraception.

Such explanations do not, however, touch on the real causes of the set of changes described - the fertility decline, increased age of marriage and numbers remaining unmarried. Underlying all these there must be something else, and the most likely candidates are a decline in the amount of land available for colonies to expand into, and a continual increase in the cost of living and hence rearing children. If this is indeed the case, the chances are that this has also changed the strongly pro-natalist Hutterite ideology, allowing for much higher levels of celibacy than before. There has also probably been a change of attitude among multiparous women to the matter of contraception.

Stress has been laid in the above account on the religious beliefs and practices of the Hutterites in explanation for their endogamy, their communal living and property-sharing, and their high reproductive rate. But this still leaves fundamental questions unanswered. Why should the beliefs be as they are? What determines such beliefs?

This question raises difficult issues. One set of answers is that provided by the history of the sect. Together with other radical Protestant

sects the Anabaptists established themselves in reaction against the establishment Protestantism of the Lutherans in the form of a return to fundamentalism,
exemplified by insistence on adult baptism.   The leaders, or "prophets" were
victimised, expelled, or executed, as were many hundreds of followers.   Not
until their arrival in America were they completely safe from persecution. Their
pro-reproductive ideology can be seen partly as a reaction against the persecutions and executions that had brought them to the brink of extinction; we do not,
however, have any information on their pre-existing reproductive rates in Europe
which may have been high.

Besides historical causes we need also to consider economic ones, in
particular land availability.   In Europe, and especially in Switzerland, land had
been in short supply throughout the Hutterites' experience; in North America
their combined resources were enough to buy large tracts of land.   The
availability of Hutterite-owned land into which to expand was important; the
fissioning of their colonies has continued until the present time.   We can almost
certainly predict that, given the same history, if the Hutterites had been as
landless after their migration as they were before, or had had limited landholdings only, they could not have developed their unique pattern of co-residence and
unlimited family size.   Had they lacked the resources to feed large families they
would necessarily have had to succumb to the pressures for smaller families that
characterised the rest of the USA from 1880 to 1950.

Given that they did obtain large landholdings on arrival in the New World,
however, a rapid rate of reproduction does not automatically follow.   Had they
been a celibate sect they could have maintained themselves by pasturing
livestock on their fields and adopting orphans into their fraternity.   Their pro-
reproductive ideology may therefore be best understood as a reaction to their
history, both in respect of ownership of land and in respect of personal factors,
for the first settlers were still very aware of the losses their families had
suffered as a result of the persecutions.

It is interesting that the sect retained its high reproductive rate in the
new, increasingly secure conditions of life in North America, unlike the rest of
the Protestant population, between 1880 and 1950.   But since it was relatively
isolated from the rest of the population, we can perhaps call on tradition, or a
kind of "cultural inertia" to explain the continuance of a pattern of mating and
reproduction more suited to a militant sect in a recovery or formative stage.
Effectively, the historical conditions of extreme uncertainty remained the
central cause not only of pro-natalist beliefs but also of the reproductive
practices of the Hutterites during the early 20th century.

Kippel in Valais

A second isolate, illustrating the opposite condition - religious rules supporting a mating pattern that leads to small family size - is found in the Swiss village of Kippel in the canton of Valais (Friedl & Ellis, 1976). This is a strongly Roman Catholic community in which late marriage coupled with a high rate of celibacy combined to limit family size and stabilize the population. Kippel is one of four communities in a valley which has altogether only some 25 square miles of agriculturally useful land. In contrast with the Hutterites, land is family-owned and because each family needs fertile (low-lying) land for crops and (higher) grazing land for animals, it has become divided up into small plots. This has been further exacerbated by the pattern of partible inheritance.

From 1900 to 1950, the population of Kippel increased from 248 to 363 (46%), partly by immigration. Marginal land was brought into use but the standard of living fell and at times there was little to eat, this situation being resolved by emigration to factories elsewhere in Valais. Late marriage and frequent celibacy were well established in the Alps before the 20th century. In the case of late marriage, restrictions were applied specifically to control family size. From 1900-1969, in Kippel, mean age at first marriage varied between 27.2 and 33.4 years for men, 26 and 29.7 years for women. Parents did not allow children to use a portion of the estate, a prerequisite for marriage, until they reached a suitable age. This was thus a secular control on reproduction.

Celibacy, however, was not a purely secular matter. According to Friedl and Ellis "there was, and still is, an extremely long religious tradition dictating chastity for both men and women. The priest literally ruled the valley." (p. 28). Valais, unlike the other Swiss cantons, was strongly Catholic, and from it the Papal Guard was, and still is, drawn. In Kippel in the 1970 census, there were 45 unmarried men and women over 40 years of age, representing 20% of all marriageable individuals. This represents the continuation of an earlier and widespread trend in Europe which was, however, nowhere more clearly marked than in the Alpine valley communities of Switzerland.

Friedl and Ellis attribute this pattern in Kippel in part to "culture lag", brought about by Kippel's isolation and the slow onset of modern influences. Second, they stress the religious tradition of celibacy, avoidance of premarital sex, and sexual restraint, accompanied by a large measure of ignorance about sex and apprehension of it, both of which tend to accompany religious taboos and secrecy about sex. They expect, however, to see a decline of the religious tradition and the acceptance of contraception in due course, as secularisation and modernisation proceed.

Although the case of Kippel is far less well documented, especially on the socio-religious side, than that of the Hutterites, the two do provide a good contrast. Both groups banned contraception. But whereas in the Hutterite religion a battery of beliefs all worked to encourage rapid and continuous procreation, in Kippel they worked just the opposite way, disapproving, delaying and disparaging sexual activity. Plentiful land was crucial for Hutterite pronatalism to succeed; in Kippel, land shortage was the material factor underlying anti-natalism. Nor did the Kippelites have the theoretical option of reversing their strategy; they were compelled to control their population because they were at the limit of the carrying capacity of the environment, given their traditional ways of exploiting it. Both isolates showed in the mid-20th century evidence of the persistence of cultural features no longer wholly appropriate. In the case of the Hutterites, their growth rate is today an anomaly, whereas at first it was an essential aspect of their recovery. In the case of the Kippelites, Friedl and Ellis state that in modern conditions the "reluctance to marry early or at all" is becoming anachronistic.

MAINSTREAM RELIGIONS

Christianity

The two examples given above serve well to illustrate the influence of religious ideas and rules on mating patterns, because we can contrast these isolates with the mainstream cultures around them. We tend to think that reproductive decisions in mainstream cultures are largely made on a secular basis. As a result, demographic analyses tend to focus on economic, educational, social class and other secular factors as underlying causes of mating patterns and fertility levels. The religious factor is, however, right under our noses in the contrast between Protestants and Roman Catholics. Despite the liberalisation and secularisation of R.C. attitudes to contraception during the 20th century (Westoff & Ryder, 1977), its central dogma remains solidly opposed to contraception. This does still affect both mating patterns and rates of reproduction in Europe and the Americas. A detailed study of all aspects of Catholic belief, attitudes, fertility and contraception was made by the U.S. Population Council in 1968 (Jones & Nortman, 1968). The authors of this study contrast Catholic doctrine on mechanical and chemical methods of birth control with that of the Protestant church which, since 1930, has given ever-increasing consent and approval to the use of all contraceptive techniques. Crude birth rates for North America and Europe indicate that Catholic fertility is essentially of the modern type, i.e. controlled, but comparisons within each country of Catholics and Protestants show higher fertility levels for Catholics in every case. In Great Britain, for example, using an index base figure of 100 for non-Catholics, the Catholic index was 122 for manual classes and 139 for non-manual (data for 1966). Catholic excess fertility was found to be from 18 to 30 percent in developed countries. They note that fundamentalist Protestant groups have higher fertility than Catholics, but these are the exceptions to the general rule of excess Catholic over Protestant fertility in developed countries.

The data show that in the developed world, Catholics use birth control methods either less efficiently or less often than Protestants. Data about efficiency are not given by Jones and Nortman (op. cit.). But data are given on wanted family size and these show that U.S. Catholics at all levels of education want as well as have a larger family than Protestants. More highly educated Catholics want and have even larger families than less educated ones, the exact opposite of the case in Protestants. The authors attribute this to "the close involvement of college-educated Catholics with their faith" even though studies have shown that college educated Catholics themselves rarely perceived their religion as underlying their desired family size.

This last point is of interest because it shows that individuals can be quite unaware of the religious basis of their ideas on family size. It also shows that religious faith is a major determinant of reproductive success at the present time.

What is the origin of the Catholic doctrines on birth control and family size? Clearly, Catholicism is much older than Protestantism, yet Protestantism itself was very pro-reproductive until the 20th century. Perhaps it is the centralisation of Catholicism, the existence of one supreme authority and the practice of successive Popes of upholding the traditions of the old church that have led to the present anomaly. Papal Encyclicals are, of course, for the Catholics living in the countries of South America as much as for Europe and North America, and in rural South America (though not in the cities) economic conditions are still like those of medieval Europe in some respects. Twelve percent of all Catholics live in rural South America, where mortality levels remain as high as they are in some African and Asian countries. Because of this, in rural South America the Papal doctrines may in some sense be "adaptive", i.e. in accord with the exigencies of the environment.

Nevertheless, in general, from a secular point of view, Catholic teaching on birth control and family size does seem anachronistic, and once again we can perhaps attribute this to "cultural lag", and predict that change will eventually come.

Protestantis n does not seem to have had such difficulty adapting to modern circumstances, no doubt because it is more recent, and also because of lack of centralisation. But there is another reason and that is the link between thrift and the accumulation of spiritual and material wealth, found in Protestantism especially (Weber, 1930, transl.). This underlying attitude may have led to decisions to curtail family size in order to maintain and increase affluence. Indeed this may be a motivating force for small family size in the affluent world still, for in affluent societies the cost of rearing an extra child is greater than in poor ones, if only because, once born, it is much more likely to survive. This question raises a major issue in the understanding of demographic transition, which can only be touched on here. This is the shifting of the idea of self-improvement from individuals themselves to their children. For instance, when parents deprive themselves financially for the sake of their children's education, we see an element of "deferred gratification" quite absent in, for example, the Hutterites, who adhere firmly to the principle that "God will provide".

Islam

Lastly, let us examine the situation in Moslem countries. Islam presents a great contrast with Christianity in respect of mating patterns and reproductive performance (Table 2). Almost all published studies of Islamic fertility rates have shown these to be very high. The early history of Islam, it is well known, was a bloody one, with bitter fighting between the followers of Mohammed and the Jews at first, then later with the Christians and indeed other Moslem factions. It is perhaps as a result of its military history that Islam has maintained a highly pro-natalist ideology. But there is another important reason, namely that the words of the Koran represent not only the writings of Mohammed and others (as does the New Testament in respect of the Apostles) but the actual words of Allah himself as revealed to Mohammed. These words, and the attitudes they enshrine, are thus wholly immutable. It is because of this that many times in the history of Islam efforts to liberalise Islamic countries have been met with civil war and a reversion to fundamentalist values. As with all major religions, doctrinal differences have divided Islamic sects against each other, but in contrast with the situation in Christianity, Islam has not seen a demographic transition along the usual lines.

Nagi (1983) showed, in a study of fertility levels in 30 Moslem countries from 1960 to 1980, that Moslem fertility remained universally high and that the economic circumstances which produced low fertility in many parts of the world had not consistently done so in Islamic countries. This was so despite the fact

Table 2: Demographic variables for 9 Moslem and 9 Protestant countries

	Religion	CBR (per 1000)	TFR	IMR (per 1000)	LEB (years)	GNP
Syria	Islam	47	7.3	57	64	1,680
Jordan	Islam	46	7.4	63	64	1,710
Saudi Arabia	Islam	42	7.2	103	56	12,180
Turkey	Islam	35	5.1	110	63	1,230
Yemen (N)	Islam	48	6.8	154	44	510
Algeria	Islam	45	7.0	109	60	2,400
Libya	Islam	46	7.2	92	58	7,500
Morocco	Islam	41	5.9	99	58	750
Tunisia	Islam	33	4.9	85	61	1,290
U.K.	Protestant	13	1.8	10.1	73	9,050
Norway	Protestant	12	1.7	7.8	76	13,820
Sweden	Protestant	11	1.6	7	76	12,400
Finland	Protestant	14	1.7	6	74	10,440
Denmark	Protestant	10	1.4	8.2	74	11,490
Netherlands	Protestant	12	1.5	8.4	76	9,910
West Germany	Protestant	10	1.3	10.1	74	11,420
Switzerland	Protestant	11	1.6	7.7	76	16,390
Iceland	Protestant	19	2.2	7.1	77	10,270

Source: World Population Data Sheet, 1985

CBR = crude birth rate; TFR = total fertility rate; IMR = infant mortality rate; LEB = life expectancy at birth; GNP = gross national product per head (US $)

that Islam does not doctrinally rule out contraception as does Catholicism. But Islam incorporates some very pronatalist features. First there is the low status of women, which tends to leave decisions about family size to the father. Second there are Koranic injunctions to be fruitful and especially to produce sons (perhaps related to early Islamic military requirements although built on a pre-existing Arab preference for sons reflecting a strongly patrilineal society). Marriage is universal, divorce and remarriage are simple matters, and women are often expected to remain at home in their role as wives and mothers, and not to seek work or other diversions. The relative seclusion of women, and traditional attitudes to them and their role in life, may be more potent influences on family size than the taboo on contraceptives is for Catholics.

There has, however, been a decline in Moslem crude birth rates and fertility rates in many countries, notably in Malaysia, Indonesia, Turkey, Tunisia and Egypt. But fertility rates for 29 Moslem countries in 1980 remained higher than the world average (3.8) or the average for the less developed countries (4.4). In sixteen of the countries included in Nagi's survey, this rate (TFR) was 6.5 or higher in 1980.

Nagi ranked 29 Moslem countries on an index of socioeconomic development. This failed to show an association between the index scores and fertility. Those countries whose fertility had declined were quite poor (e.g. Bangladesh), while for Jordan and Libya socioeconomic progress was associated with largely unchanged high fertility levels. Infant mortality rates were mostly high, but low rates were not always associated with lowered fertility, e.g. in Jordan. The only factors found by Nagi to have consistently reduced fertility levels were the existence and effectiveness of government family planning programmes, in Indonesia, Iran, Malaysia and Tunisia. Later age at marriage was another factor, but not consistently so (see also Mauldin & Berelson, 1978, for similar findings).

An even more recent study is that of Ahmad (1985) on factors affecting fertility in four Moslem populations (Bangladesh, Java, Jordan and Pakistan). This has confirmed that fertility transition is not yet occurring in these countries. Variations in fertility were found to be explained by four main factors: age at first marriage, duration of marriage, stability of first marriage and experience of child loss. As regards educational level, he found that the effect of increasing level of education on husband or wife tended to favour higher fertility in the four countries concerned.

This was also found in U.S. Catholics (see above) where it was attributed to greater adherence to doctrine. But the same conclusion cannot necessarily be drawn for Islam because higher educational level might well, in this case, go with improved health and reduced infant mortality. It does show, however, that an ethic of small family size is not a feature of more educated and doubtless more affluent Moslems.

The historical emphasis placed by Islam on high fertility, together with the traditionally low status of Moslem women, may well be influential in maintaining high fertility rates today. But we need to consider also the material conditions of existence which we have seen are closely linked to fertility. Historically, and at the present time, the countries of Asia, the Middle and Near East, have differed from Europe in one major respect, according to Jones (1981). This has always been the pattern of distribution of wealth. Whereas in Europe this pattern has become progressively more egalitarian, in the former areas it has not, and wealth has always been in the hands of a few, while the vast majority has remained very poor. Such accepted differentials are reflected at the family level in the Moslem institution of polygyny. This marriage pattern very much antedates Islam, but was strongly favoured by Mohammed after the battle of Badr, and has persisted as an institution well suited to a society with major wealth differentials. Rich men can afford many wives and the reproductive success of women is best assured with these rich men; the poor cannot afford any wives at all, nor would women be well supported by them. Polygyny enables many children to be born and supported in a poor community in which a few men are rich.

Comparing Europe with the East, one cannot but be struck by the contrast in environments. Russell (1967) and Jones (1981) have both contrasted the poor, dry environment of Western Asia with the wet rich lands of Western Europe. The disease environment of Islamic countries has been far more hostile than that of Europe, with far greater famines and plagues than Europe has ever known, and a consistently higher level of endemic infectious diseases, persisting to the present time.

Nagi (op. cit.) was unable to correlate fertility with modernisation in his survey of 29 Moslem countries worldwide. However, such a diverse sample presents many problems. If demographic variables for nine Moslem countries from the Moslem heartland - the Middle East, Near East and North Africa, representing a fairly homogeneous, hot, dry zone - are compared with nine Protestant countries from Western Europe and Scandinavia, representing a much cooler, wetter environment, the contrasts and the resulting correlations become very evident indeed (Table 2). Fertility in these two areas provides a great contrast. It has been shown (e.g. by Mauldin & Berelson, 1978) that fertility

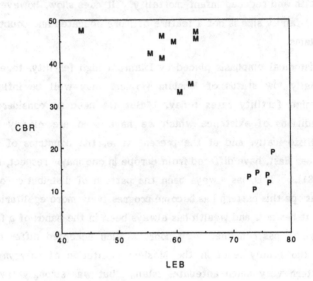

Figure 2:   Comparison of life expectancy and crude birth rates in 9 Moslem and 9 Protestant countries.
M = Moslem;   P = Protestant;   LEB = life expectancy at birth;   CBR = crude birth rate.
(From the 1985 World Population Data Sheet.)

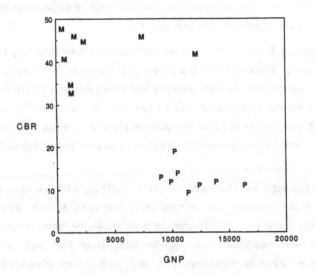

Figure 3:   Income (GNP per head) and crude birth rate in 9 Moslem and 9 Protestant countries.
M = Moslem;   P = Protestant;   GNP = gross national product per head (US$);   CBR = crude birth rate.
(From the 1985 World Population Data Sheet.)

differentials are closely related to mortality rates and to differences in standards of living, as shown for instance by the figures for life expectancy (Figure 2) or for per head GNP (Figure 3). This contrast and the others illustrated by the data in Table 2 show that at this level there are two quite distinct reproductive strategies at work in these two different physical and socio-religious environments.

In the case of North Africa, the Near and Middle East, the whole zone is characterised by levels of both mortality and fertility that are high in absolute terms. Crude birth rates are higher than 30 per thousand population, infant mortality rates are in all cases higher than 50 per thousand births, maximum average life expectancy at birth is 64 years. Figures for GNP per head are misleading in the cases of Saudi Arabia and Libya, where they are inflated by oil revenues to the countries concerned. In other cases they do not exceed $2,400 per person per annum.

Comparison with north-west Europe shows a complete absence of overlap for these quantities. The highest crude birth rate is that of Iceland, at 19 per thousand population. The highest infant mortality rate is 10.1 per thousand births. Life expectancy in all cases exceeds 70 years, and the figures for GNP per head are realistic, and do not fall below $9,000 per annum.

It has been argued elsewhere that the religions associated with these two contrasting demographic situations are closely related to two sets of biological pressures (Reynolds & Tanner, 1983). Islam is far more pronatalist than Protestant Christianity. To that extent, the religons are environmentally adapted, and underlie local demographic regimes.

Essentially the same argument can be made for the Hutterites and the people of Kippel. Data exist for crude birth rates, total fertility rates and infant mortality rates for the Hutterites at two dates (1950 and 1971), and for the people of Kippel in 1970. If we now place these isolates on relevant figures for fertility and mortality in the nine Moslem and nine Protestant countries selected previously, we can see how they compare (Figures 4 and 5). In the case of the Hutterites, the data for both times place them well to the left of the regression line for the Moslem-Protestant data. They thus show a considerable excess of fertility in relation to mortality. This excess has been maintained between the two dates. The later sample, compared with the former, has followed a trajectory parallel with the Moslem-Protestant line and consistently to the left of it.

The position of Kippel is perhaps notable because we might have expected a lower fertility level than the actual one, given the circumstances described.

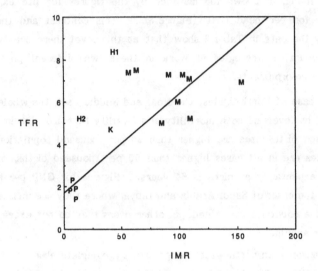

Figure 4:   Infant mortality rates and total fertility rates in 9 Moslem and 9
Protestant countries, showing the position of the Hutterites (1950
and 1971) and Kippel (1970) (based on Moslem and Protestant
countries only in order to indicate the comparative positions of the
isolates).
M = Moslem;   P = Protestant;   H1 = Hutterites (1950);   H2 =
Hutterites (1971);   K = Kippel (1970);   IMR = infant mortality rate.
TFR = total fertility rate.

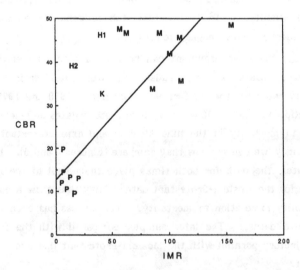

Figure 5:   Infant inortality rates and crude birth rates for countries and
populations (and abbreviations) as in Figure 4.

The relatively high fertility of Kippel indicates that despite the emphasis on late marriage, and high celibacy levels, fertility control was not effective, and this is supported by the data on population increase and the reported high level of emigration. Only contraception, it would seem, can reduce this fertility, and this may now be happening; if so there will need to be a readjustment of religious feelings about this, at the personal if not at the official level.

The relation between these environmental conditions and the religious rules, mating patterns and patterns of fertility of countries today are surely exceedingly complex. But these factors are just as surely related to each other. This paper has tried to show that religious ideas and rules need to be considered just as carefully as the more often measured secular factors in understanding the biology of present-day populations.

REFERENCES

Ahmad, S. (1985).   Factors affecting fertility in four Moslim populations; a multivariate analysis. J.Biosoc.Sci., **17**, 305-316.

Bateson, P.P.G. (1982). Preferences for cousins in Japanese quail. Nature, **295**, 236-7.

Dumont, L. (1970). Homo hierarchicus. London: Weidenfeld & Nicholson.

Eaton, J.W. & Mayer, A.J. (1953).   The social biology of very high fertility among the Hutterites. Human Biology, **25**, 206-264.

Friedl, J. & Ellis, W.S. (1976).   Celibacy, late marriage and potential mates in a Swiss isolate. Human Biology, **48**, 23-35.

Hostetler, J. (1974). Communitarian Societies. New York: Holt, Rinehart & Winston.

Jones, E.L. (1981).   The European Miracle. Cambridge: Cambridge University Press.

Jones, J. & Nortman, D. (1968).   Roman Catholic fertility and family planning. Studies in Family Planning, **34**, 1-27.

Laing, L.M. (1980).   Declining fertility in a religious isolate: the Hutterite population of Alberta, Canada, 1951-1971. Human Biology, **52**, 288-310.

Lang, H. & Goehlen, R. (1985).   Completed fertility of the Hutterites: a revision. Current Anthropology, **26**, 395.

Mauldin, W.P. & Berelson, B. (1978). . Conditions of fertility decline in developing countries, 1965-75. Studies in Family Planning, **9**, 89-147.

McKusick, V.A. (1973).   Genetic studies in American inbred populations with particular reference to the Old Order Amish. Israeli J.Med.Science, **9**, 1276-84.

Nagi, M.H. (1983).   Trends in Moslem fertility and the application of the demographic transition model. Social Biology, **30**, 245-262.

Reynolds, V. & Tanner, R. (1983). The Biology of Religion. Harlow: Longman.

Russell, W.M.S. (1967). Man, Nature and History. London: Aldus.

Schull, W.J. (1953).   The effect of Christianity on consanguinity in Nagasaki. American Anthropologist, 55, 74-88.

Weber, M. (1930, transl.).   The Protestant Ethic and the Spirit of Capitalism. London: Unwin.

Westoff, C.F. & Ryder, N.B. (1977).   The Contraceptive Revolution.   Princeton: Princeton University Press.

THE RELEVANCE OF THE
POLYGYNY THRESHOLD MODEL TO HUMANS

M. BORGERHOFF MULDER

Department of Anthropology, Northwestern University,
Evanston, Illinois, U.S.A.

and the Large Animal Research Group
Department of Zoology, University of Cambridge, Cambridge, England

INTRODUCTION

The principal explanation proposed for the evolution of polygynous mating in non-human species is the differential ability of males to defend the resources on which females depend for survivorship and reproduction. In such "resource defence" polygynous mating systems (Emlen & Oring, 1977) females are thought to select a mate according to the quality of resources he can offer her. The difference in the resources held by males required to make polygynous mating advantageous to females is called the "polygyny threshold" (Verner & Willson, 1966; Orians, 1969). Male red-winged blackbirds, for example, hold territories of varying qualities (Lenington, 1980), and the number of females a male can attract is determined, in part, by the quality of his territory. Variation among males in their ability to monopolise resources predicts the incidence of polygynous mating in many species of birds (see Wittenberger, 1981) and in some non-human mammals (Downhower & Armitage, 1971; Elliott, 1975).

In humans, correlations between the extent of differentiation in resource-holding among men, in particular socioeconomic stratification within a society, and the degree of overall polygyny within that society are now well attested cross-culturally (e.g. Osmand, 1969; Clignet, 1970; Blumberg & Winch, 1972; Spencer, 1980; Betzig, 1986). Furthermore, there is extensive evidence that within societies wealthy men accumulate more wives than their less wealthy competitors (Mair, 1969; Clignet, 1979; Irons, 1979; Wilson, 1981; Borgerhoff Mulder, 1987a).

These observations have led some authors to suggest that female preferences for wealthy men may be an important factor in the evolution and current incidence of polygyny (van den Berghe, 1969; Irons, 1983; Hartung, 1982). Nevertheless, the precise mechanisms whereby wealthy men achieve higher mating success than others have not been examined quantitatively. This lack of detailed information has rightly led other authors to question whether men are selectively chosen by women on the basis of the resources they can offer, and whether or not the polygyny threshold model is indeed applicable to humans (Scott, 1982; Daly & Wilson, 1983; Gray, 1985; Flinn & Low, 1986).

To demonstrate the utility of the polygyny threshold model in human societies, precise data on the nature of mating competition and the causes of mating success in specific ethnographic contexts are required that show (i) wealthy men acquire more wives than poorer men, and (ii) their marital success is mediated through the preference of women, or their kin, for wealthy husbands. This study presents relevant data from the Kipsigis of Kenya.

ETHNOGRAPHIC BACKGROUND

The Kipsigis are a Nilo-Hamitic people living in Kericho District in south-western Kenya. They are part of the Kalenjin language group, and are thought to have moved into their present area in the late eighteenth century (Orchardson, 1961). Until the 1930's they were largely pastoral, keeping cattle, sheep and goats, but after the arrival of the Europeans in the area, began to adopt maize as a subsistence, and later, cash crop. This development led to the emergence of individual land holdings which, like livestock, are inherited from father to son (Manners, 1967).

While individual land titles were established over 30 years ago in some parts of Kericho District, other parts were still being adjudicated during the period of fieldwork (Borgerhoff Mulder, 1985). Notwithstanding these differences in legality of tenure, each man inherits recognised rights to a specific piece of land, many of which are now fenced as plots. Some men buy second plots. Some plots are large (maximum in this sample 300 acres), but much of the land in the area is rough and stony, while other parts are poorly drained. With mounting population pressure, land is becoming an increasingly critical resource (e.g. Manners, 1967), although cattle retain economic importance, both for dairy produce and as a currency for bridewealth payments.

All Kipsigis men belong to an *ipinda* (age set), membership of which is determined by year of circumcision (Prins, 1953). The three such cohorts who had completed, or almost completed, their marital careers (Borgerhoff Mulder, 1988a) on which this paper focuses are the *Nyongi* who were circumcised before 1922, the *Maina* (1922-1930) and the *Chuma I* (1931-38).

Boys in these cohorts were circumcised at the mean age of 15.7 (range 11-19 years) and married at 26.1 (18-46). First marriages are arranged by parents of the prospective bride and groom, with the groom's father initiating marriage negotiations by making a bridewealth offer, consisting of cows, goats, sheep and cash to the father of the potential bride (Borgerhoff Mulder, 1988b). Much negotiation over the bridewealth takes place, and only after several months of deliberation and bargaining do the bride's parents, together with their close kin, decide between the competing offers made for their daughter. Payments and negotiations for subsequent marriages are, by contrast, the responsibility of the groom himself.

Payment of bridewealth procures a man's rights to his wife's children (Peristiany, 1939) and to her labour services. Marriage is highly stable, and divorce almost unknown (Peristiany, 1939; Orchardson, 1961; Borgerhoff Mulder, 1988b), although a dissatisfied wife will run home to her parents on a temporary or relatively permanent basis if she feels mistreated in her marital home. Furthermore, the incidence of married women bearing illegitimate children is extremely low, with only 0.8% of all children in the present study reportedly *not* fathered by their mother's husband (Borgerhoff Mulder, 1987a).

Some men reaching adulthood fail to marry, usually for reported reasons of poverty or individual whim. They often leave the community to seek wage labour on the tea plantations near Kericho, where they may marry later in life.

METHODS

The study site of 35 square kilometres incorporates several *kokwotinwek* (sing. *kokwet*, neighbourhoods) in the south-western part of Kericho District, bordering Kisii and Narok Districts.

A census of all households in the study area was made between June 1982 and December 1983. Questioning each individual in kipsigis, the dates of his/her birth, circumcision and marriages, and the numbers of livebirths and surviving offspring, were determined. Discrepancies between reports of husbands and wives were subsequently re-checked with both parties, and amended accordingly. Life history information on polygynously-married women non-resident in the study area were taken from their husbands, and verified with other co-wives or relatives.

Past life history events could be accurately dated to the year by cross-referencing to male circumcision ceremonies, severe droughts and other events of known date (c.f. Central Bureau of Statistics, 1980). Numbers of offspring born to men of the *Nyongi*, *Maina* and *Chuma I* cohorts are considered as good estimates of the lifetime reproductive success of these men (Borgerhoff Mulder, 1988a).

The capital assets of a man, his land, cattle, sheep, goats, chickens and other items of household equipment were noted in the census (Borgerhoff Mulder, 1987a). Many of these measures were highly intercorrelated, and the best predictor of wealth differences between men was the number of acres held. Little land has been bought or sold since the 1930's, so the size of a man's landholding in 1982-83 provides a relatively robust measure of wealth differences between men throughout their reproductive careers. It was not possible to measure wealth differences among *Nyongi* men because they had already completed the process of dividing their land and cattle among their sons. Although in practice the process of inheritance had occurred in the *Maina* and, to a lesser extent, in the *Chuma I* cohorts, these men were still recognised legal owners.

Parents were interviewed, mothers (n=66) and fathers (n=28) separately, about the desired qualities in a potential husband for their daughters. These interviews were conducted, again in kipsigis, with parents of all ages (not just those in the *Nyongi*, *Maina* and *Chuma I* cohorts), but were restricted to families in a subsample (4 *kokwotinwek*) among whom intensive behavioral observations were conducted. Rapport with these families was particularly high, and in most cases my initial questions led to long discussions of what Kipsigis look for in their daughters' spouses. During these conversations it was noted whether the following qualities in a groom were positively or negatively evaluated, by explicitly initiating discussion on each quality and then recording the interviewee's response, more or less verbatim: (i) wealth (*ne tinye mogornotet*; (ii) respect (*ne tinye tegisto*);   (iii) education (*ne kisomani*);   (iv) industriousness (*ne boisie komie*);   (v) character (*ne kararan*).   The fathers and mothers also were asked directly about their general opinions on polygynous marriage, for themselves, their offspring and for people in the community at large.

CAUSES OF MATING SUCCESS AMONG KIPSIGIS MEN

Differences in lifetime reproductive success between men can be attributed largely to the practice of polygynous marriage among Kipsigis. A model that determines the contribution of different components of reproduction to overall reproductive variance (Brown, 1988) shows that the mean number of non-menopausal wives married to a man throughout his marital lifespan accounts for between 46 and 71% of variance in reproductive success among men (Table 1).

Table 1:   Proportional contributions to variation in lifetime reproductive success

	Lifespan	Married Lifespan	Polygyny	Offspring
Nyongi (before 1922) n = 29	8.2	9.3	46.1	35.7
Maina (1922-1930) n = 40	2.4	4.2	50.5	9.5
Chuma I (1931-1938) n = 38	1.0	9.4	71.0	34.8

Lifespan: Date of circumcision could not always be determined; where age at circumcision was known, lifespan could be calculated directly; where not, it was estimated by assuming a man was circumcised at the median age for that cohort.

Married lifespan: The proportion of lifespan spent married was calculated as the years elapsed since a man's first marriage, divided by lifespan.

Polygyny: Mean number of reproductive wives per married year was calculated by dividing total reproductive wife-years by married lifespan, thus providing a measure of polygyny.

Offspring: The number of surviving offspring per reproductive wife-year was calculated as the number of surviving offspring divided by the total number of reproductive wife-years.

Lifetime reproductive success is a product of these four components. The proportion contributions to variation in lifetime reproductive success are calculated according to Brown (1988). On account of small negative covariations between components (see Borgerhoff Mulder, 1988b), percentages do not sum exactly to 100%. (From Borgerhoff Mulder, 1987a.)

From what does this difference in the ability of men to acquire wives stem? The major correlate of reproductive variance among men is their wealth (Borgerhoff Mulder, 1987a), and this is due to the strong association between the wealth of a man and the number of his wives (Table 2). Differences in education and employment among men do not account for these associations (see Borgerhoff Mulder, 1987a).

Table 2:   Pearson correlation coefficients between wealth and polygyny
               across cohort

Cohort	Polygyny
Maina (1922-1930) n=25	.91***
Chuma I (1931-1938) n=37	.49***

*** denotes p <0.001

Table 3:   Percent of sample of parents who view wealth, respect,
               industriousness, education and character as preferred qualities in a
               potential son-in-law

	All sample (n=94)	Fathers (n=28)	Mothers (n=66)
Wealth	54	29	65
Respect	61	71	56
Industriousness	9	4	12
Education	41	14	53
Character	17	7	21

Percentages sum to over 100 because any parent could name more
than one quality in a potential son-in-law.

It is highly probable that wealth differences among men are the cause of
differences in the numbers of their wives.   While payments for first marriages
are the responsibility of a man's father, second and subsequent marriage
payments must be raised by the prospective groom himself.   Given that the
mean bridewealth for marriages occurring after 1959 is 6 cows, 6 goats and 800
Kenyan shillings, constituting approximately one third of an average man's
capital assets (Borgerhoff Mulder, 1988b), differences in the availability of
resources among men are likely to be a determinant of their ability to take
second and subsequent wives.

Interview data support this argument.   Fifty four percent (29% of all
fathers and 65% of all mothers interviewed) considered wealth as an important
quality of a potential son-in-law (Table 3).   While respect was favoured even
more often (61% of all interviews), wealth and respect are highly correlated

among the Kipsigis (shown through the use of informant ranking techniques). In 39% of all interviews, parents spontaneously suggested that their daughter would be less likely to suffer in her marital home if she was married to a rich man, generally adding that she would be less likely to run home and that they, as parents, would not have to bear the costs of looking after her children.

Fathers were apparently less concerned with the wealth of a potential son-in-law than were mothers (29% versus 65%, $\chi^2 = 10.60$, df = 1, p = 0.001). Commonly fathers explained that inherited wealth, at least in its traditional form as cattle, is easily lost (through theft, sickness, or ill fortune), and that therefore the best indicator of a man's ability to provide is his respect (71%), arguing that a well-respected man will not be left to suffer by neighbours and relatives, even if his cattle die. Younger men in the interview sample, aged less than 50 (n = 16), who increasingly think of wealth in terms of land rather than cattle since the settlement of Kipsigis on plots, were more inclined to mention the wealth of a prospective son-in-law as an important quality (63%) than were older men (20%) ($\chi^2 = 4.1$, d.f. = 1, p < 0.05); this suggests that as resources become more stable and predictable over time, wealth differences among prospective grooms are considered more crucial.

The only other quality in a prospective son-in-law to be mentioned as important in more than 20% of interviews was education (41%). Education was particularly favoured among parents aged less than 50 years, with 60% of the sample responding positively to the suggestion that a son-in-law should be educated. In the most recent cohort (circumcised between 1964-1978), men who have completed over six years of education are more likely to obtain employment ($\chi^2 = 25.05$, n = 281, p < 0.001), and employment can provide a critical supplement to the farm income. Favouring education may therefore be congruent with favouring wealth in a prospective son-in-law.

Directly questioned about polygyny, 86% of mothers approved of it generally and 67% favoured it for their children; 40% spontaneously added that polygynous marriage is only acceptable if the husband is rich. All but one father generally approved of polygyny, with 64% spontaneously stating that economic problems and co-wife strife arise if insufficient resources are available to support each wife.

Although only suggestive, these interviews indicate that polygyny, albeit deeply engrained in the Kipsigis way of life, is contingent on the economic circumstances of the groom. These attitudinal data suggest that parents may selectively choose, from a number of prospective sons-in-law making bridewealth offers, a man with adequate resources, irrespective of his current marital status.

It can therefore be concluded that variation in the number of a Kipsigis man's wives is principally due to wealth differences, and that this is in part a consequence of parental preferences to give their daughters to men with sufficient resources.

THE NATURE OF REPRODUCTIVE COMPETITION IN KIPSIGIS MEN

Variation in numbers of surviving offspring for all men residing in the study area belonging to the three oldest cohorts are shown in Figure 1.   The means and variances for the *Nyongi*, *Maina* and *Chuma I* cohorts respectively are 11.07 children (54.93), 14.47 (164.93) and 10.71 (36.26).   The mean values and variances in completed family size and three cohorts of comparably-aged women are 4.6 (5.85), 4.8 (7.11) and 6.8 (7.84) (Borgerhoff Mulder, 1988a).   Dividing the standardised variance (χ^2/o^2) in completed family size for men in each cohort by the standardised variance for the equivalent cohort of women gives indices of the comparative intensity of selection (Im/If;   Wade & Arnold, 1980) of 1.61 (*Nyongi*), 2.63 (*Maina*) and 1.88 (*Chuma I*).

The comparative intensity of selection, as measured in this study, is likely to be an underestimate.   Many of the men who leave the community and live on tea plantations in the north, supporting themselves on wage labour rather than the family plot, fail to marry.   Restricting the sample to men who marry and stay in the area biases the sample towards successful men.   Were it possible to include non-reproductive emigrants in the analysis, the comparative intensity of selection calculated for these three cohorts might be higher.   Even without such a correction, it is clear that variance in lifetime reproductive success of men greatly exceeds that of women in all three cohorts for which data on lifetime reproductive success are available.

The number of surviving offspring born per year to men in the *Nyongi* cohort are plotted for 5-year intervals since circumcision (Figure 2);   for comparative purposes the number of surviving offspring is also plotted for women (for 5-year intervals since marriage).   The pattern of age-specific breeding success in men is similar in all three cohorts.   Two points should be noted from Figure 2.   First, the reproductive careers of men extend to beyond 60 years after circumcision, and are considerably longer than those of women. Second, the rate of surviving offspring born per year is highest beween 20-25 years after men's circumcision, when the man is nearing 40.

The weak effects of age on breeding success in males can be attributed to the finding that acquiring wives among Kipsigis is not strongly age-dependent.

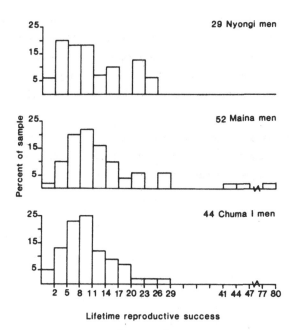

Figure 1: Distributions of lifetime reproductive success of 29 *Nyongi* (circumcised before 1922), 52 *Maina* (1922-30) and 44 *Chuma I* (1931-38) men.

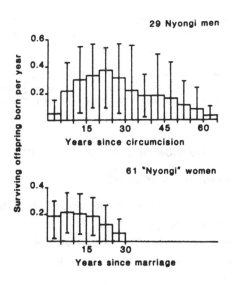

Figure 2: Surviving offspring born per year for 29 Nyongi men by 5-year intervals since circumcision (median circumcision date 1914) compared with the number of surviving offspring born per year for 61 "Nyongi" women by 5-year intervals since marriage (married before 1930) (from Borgerhoff Mulder, 1988a).

Plotting the proportion of men who remarry in each 5-year interval since first marriage shows that while subsequent marriages peak between 4 and 24 years of first marriage, 37% of all subsequent marriages by men in both the *Maina* and *Chuma I* cohorts are contracted after 24 years of marriage, and that some men in the *Maina* cohort continue to marry over 54 years after their first marriage (Figure 3).

In sum, these analyses show that variance in the lifetime reproductive success of men exceeds that of women, and that men's reproductive careers are both longer and less age-dependent than those of women.

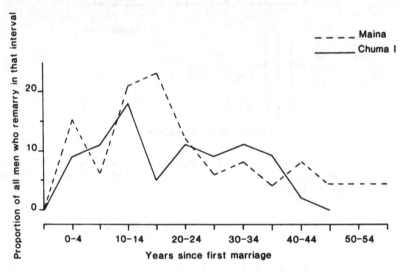

Figure 3:   The age-specific incidence of remarriage is shown by plotting the proportion of men in the Maina and Chuma I cohorts who remarry in each 5-year interval since their first marriage.

DISCUSSION OF KIPSIGIS MARRIAGE

Relevance of the polygyny threshold model

Polygynous marriage is shown in the Kipsigis study to be closely contingent on economic success, particularly resource ownership, with wealth differences among men accounting for as much as 83% of the variance in the mean number of reproductive wives married to a *Maina* man during his married lifespan.

The strong association between wealth and polygyny found in the Kipsigis must result, in part, from the custom of bridewealth payments; payments for secondary (and subsequent) marriages are entirely the responsibility of the groom, and for many men this expense makes polygyny impossible. This suggests that differential access to resources is not only a correlate but a major determinant of polygyny, at least in societies such as the Kipsigis where bridewealth is paid. The association between wealth and polygyny may result, in part, from the tendency of Kipsigis parents to favour wealthy husbands for their daughters. The offers of poor men who aspire to polygyny might be more frequently rejected, if the attitudinal data presented here bear any relation to the way in which parents chose between prospective suitors.

Despite the apparent relevance of the polygyny threshold model to this case, to what extent are the critical assumptions of this model actually met in the Kipsigis? The two key assumptions of the model are as follows (Ralls, 1977): (i) male parental investment is important to offspring survival and reproduction, characteristically with offspring living on their father's territory and being at least partially dependent on his resources (if this were not so, the prospect of having to share a male's parental investment would be unlikely to be the factor deterring females from mating polygynously); (ii) females are free to choose between males in such a way as to maximise their reproductive success – how these conditions, although violated in some mammalian species (cf. Ralls, 1977) are met in human societies such as the Kipsigis, is now discussed.

Like most pastoral people, Kipsigis are strongly patrilineally organised, and practise virilocal post-marital residence. Most importantly, rights to the resources critical to survival and reproduction are held and inherited exclusively by men. Although the labour contributions of women are high (Borgerhoff Mulder & Caro, 1985), women cannot be productive without access to land or cattle. Women depend entirely first on their fathers, then on their husbands, for these resources, with the result that offspring survivorship among Kipsigis is shown to be affected by the wealth of a child's father (Borgerhoff Mulder, 1987b). The first assumption of the polygyny threshold model is therefore met.

As regards the second assumption, it is the girl's parents who choose her husband. Although the reproductive interests of parents and offspring may overlap (Hamilton, 1964), parents might be expected in some situations to choose a husband for their daughter with aims other than that of enhancing her reproductive success. For example, they may try, through marriage, to forge alliances between families (as among Bedouin Arabs; Peters, 1980), to gain access to political office (as among the Tshidi of southern Africa; Comaroff & Comaroff, 1980), or to settle disputes (as in the Nuer of Sudan; Evans Pritchard, 1940). While such motivations cannot be ruled out in the Kipsigis case, Kipsigis do not appear to manipulate bridewealth offers and demands in such a way as to attract affines richer or more powerful than themselves (Borgerhoff Mulder, 1988b). Furthermore, the interview responses reported in this chapter clearly indicate that parents are concerned about the ability of a prospective son-in-law to support his wife, or wives. In the Kipsigis case then, parental attitudes concerning potential sons-in-law appear to reflect their daughter's interests, and the second assumption of the polygyny threshold model is apparently not violated.

Resource defence polygyny and age-specific reproduction

The type of competition between males implicated in resource defence polygyny, where females selectively choose among males on the basis of their resources, may in part contribute to the weak effects of age on reproductive rates in Kipsigis men. In most non-human mammals and birds, breeding success is strongly dependent on age, and male reproductive careers are markedly shorter than those of females, e.g. red deer (Clutton-Brock, 1983), elephant seals (Le Boeuf & Reiter, 1988), black grouse (Kruijt & Vos, 1987), lions (Bygott et al, 1979). In red deer and elephant seals, males compete directly over access to females and then defend these harems against other males (Clutton-Brock, Guinness & Albon, 1982; Le Boeuf, 1974); such systems are termed "harem defence" polygyny (Emlen & Oring, 1977). In black grouse, males compete directly with one another (at leks) and are then chosen by females (Vos, 1983). These observations suggest that the highly age-dependent and relatively short reproductive careers of the more competitive sex (males) may be attributable to the high physical costs of competition over mates (Clutton-Brock, 1983, 1988). What is there about the form of mate competition among Kipsigis men that leads to weaker effects of age on breeding success than observed in the species described above?

The explanation lies perhaps in three aspects of resource defence polygyny that reduce the costs of competition to males. First, Kipsigis men compete as coalitions, not as individuals. A man is foremost a member of an extended family, principally his patrilineage, and can call on the assistance of his brothers, sons, uncles and other close kin in raiding cattle from other tribes, in negotiations over land and livestock, and in other disputes (Peristiany, 1939). If injured or ill, he can rely on his kin to defend his interests; if temporarily destitute, he and his family are helped by relatives, and even neighbours (Peristiany, 1939). Benefits of coalitions are reported in lions (Bygott et al, 1979; Packer et al, 1988) where larger coalitions of males achieve longer tenure of prides and therefore enjoy longer reproductive lifespans, and also in cheetahs where coalitions of males may have enhanced survivorship (van de Werken, 1968; Caro & Collins, 1987).

Second, the costs of competition are reduced by prohibition on the use of violence among Kipsigis. Although armed attacks on Masai, Kisii and Luo for land and livestock were common in the past (Peristiany, 1939; Orchardson, 1961), and cattle raids against Masai and Kisii still occur (despite their illegality), disputes among Kipsigis are settled primarily without recourse to violence. Clashes of interest over rights or resources are judged by a council of old men (or occasionally nowadays taken to the national judicial courts). Where competition takes the form of debate, and success is dependent on influence and respect, both of which are highly correlated with age, the survivorship costs of competition are probably lower than where competition takes the form of direct physical conflict. The high physiological costs of intense male competition (Clutton-Brock et al, 1985), and certainly the direct costs of physical contests (Thornhill & Alcock, 1983), are probably reduced when disputes are settled through signals and displays. The relatively non-aggressive dispute settlement characterising many human societies such as the Kipsigis may therefore in part account for the long reproductive lifespan of males and the weak effect of age on male breeding success, in comparison with males of other polygynous species where conventions on restricting violence are less elaborate.

A third possible explanation for the apparently low costs of polygyny to males is that in polygynous systems where males compete over and monopolise resources, these resources themselves may serve as buffers against physical deterioration, guaranteeing food and income, particularly in times of drought or other stress. There is, as yet, no clear evidence that in non-human species, resource defence polygyny is associated with weaker effects of age on male breeding success than in harem defence polygyny; furthermore such an association would only be predicted where males hold resources that are critical to their own, as well as to the female's survival, and where resources are held for long periods of time. Nevertheless, there is a possibility that the form that male competition takes, namely whether resources or females are defended, may be associated with differences in age-specific breeding success and length of reproductive career, both across species and societies.

In summary, examination of the associations between patterns of competition between males over females and both the effects of age on breeding success and the relative lengths of male and female breeding periods are as yet insufficiently detailed to determine why the human pattern deviates from that

observed in non-human polygynous species.   Furthermore there has been no
systematic attempt to link age-specific parameters of reproduction to the
particular pattern of mating competition among males in different human
societies.   Nevertheless, as a tentative conclusion, it can be suggested that the
pattern of competition over wives observed among the Kipsigis, where valuable
resources such as land and cattle are held by coalitions of kin and where there
are rigourously observed sanctions against violence, may have promoted the
relatively weak effects of age on male breeding success.

GENERAL DISCUSSION

Looking more generally at the application of the polygyny threshold model
to human societies, numerous ethnographic references indicate that economically or socially prominent members of societies tend to accumulate larger
numbers of wives than do commoners (e.g. Goldschmidt & Kunkel, 1971; Yalman,
1971; Chagnon, 1979a; Irons, 1979a; Faux & Miller, 1984; Hill, 1984; Betzig,
1986).   Nevertheless, application of the polygyny threshold model to human
marriage has been questioned on two counts (Daly & Wilson, 1986; Gray, 1985).
First, are the assumptions of the model (male parental investment and female
choice) generally met in human societies?   Secondly, are the predictions
concerning the equal reproductive success of monogamously and polygynously
married women confirmed (Verner, 1964; Verner & Willson, 1966; Orians, 1969)?

The predominant practice of inheritance from fathers to sons (Hartung,
1976, 1982) suggests that most of the resources critical to survival and
reproduction are generally held by men.   If the coarse codings of the
Ethnographic Atlas are accepted, women and offspring are at least in part
dependent on men in most societies and the first assumption of the model is
therefore met.   Evidence for the second assumption, that females are free to
choose their spouses, is less clear.   In many societies, it is parents who select
husbands for their daughters, raising the suggestion (discussed above) that the
choice of spouse reflects parents' manipulative political interests rather than the
reproductive interests of their daughters, and that the polygyny threshold model
is therefore inappropriate (Daly & Wilson, 1983; Gray, 1985).   While most
anthropologists, following their traditional interest in kinship (e.g. Fortes, 1962),
stress the social and political aspects of marriage, the Kipsigis data show that
economic factors affecting a daughter's welfare can influence parental decisions
concerning preferred qualities in sons-in-law.   These findings question the
general view that marriages arranged by parents will necessarily fail to reflect
daughters' interests.   Nevertheless, there are other cases where daughters are

clearly the victims of parents' machinations, with parents prejudicing their daughters' reproductive interests in order to maximise the numbers of their own grandchildren, sometimes sacrificing their daughters' reproductive interests through measures as extreme as selective female infanticide (Dickemann, 1979). In sum, on current evidence it is unclear to what extent the sacrifice of daughters' reproductive interests is general across human societies, and therefore whether the condition of female choice is usually met. Detailed studies of parental attitudes may nevertheless reveal that daughters' interests are of more importance than previously thought.

The second reason why investigators have been reluctant to accept the validity of the polygyny threshold model in humans is that they have been unable to establish that polygynously-married women achieve more, or at least as many, children as they would have if married monogamously to a male with fewer resources. Both theoretical (Altmann, Wagner & Lenington, 1977) and empirical (Borgerhoff, Mulder & Caro, 1983) problems arise in testing this prediction of the model. Most studies of polygyny to date show decreased fertility of polygynously-married women (e.g. Smith & Kunz, 1976; but see Borgerhoff Mulder, 1988a). Nevertheless, problems arise in determining whether women married polygynously attain low fertility for other reasons, perhaps their age at marriage or their socio-economic background. In brief, it is premature to dismiss the applicability of the polygyny threshold hypothesis to humans because of the failure to establish equal reproductive success of monogamously and polygynously married women, in so far as adequate studies of the effects of marital status on reproductive success have not yet been conducted (cf. Irons, 1983).

If it is accepted that polygyny is contingent on not only economic differences among men but also preferences among women and their kin for husbands who can support children, some provisional suggestions can be made as to the kinds of social and ecological factors that will determine whether or not polygyny is a feasible option for either sex: for example, the ease with which wealth can be monopolised by men, the extent to which wealth differences between men can be maintained over time, the ability of women to attain independent access to resources, and the effects of resources on the survivorship of women and their offspring.

Other forms of competition for wives: ecological constraints

Polygynous marriage is not always simply attributable to the differential ability of males to monopolise or defend resources. In such cases, polygyny cannot result from female preferences for wealthy males and the polygyny threshold model is clearly not applicable. These cases indicate the complexity of patterns of male competition, even among societies nominally characterised as "polygynous", and they highlight the need for a systematic explanation for different patterns of inter-male competition, examining the interrelations between ecological factors and the reproductive strategies of males and females.

In conclusion, a brief discussion follows of some cases where polygyny clearly is not contingent on wealth, and others where explanations for the incidence of polygyny have focussed on factors other than resource differentials and female choice.

First, there is evidence that some of the polygynous mating that occurs in prescriptively monogamous societies, in the form of rape, prostitution and casual affairs (c.f. Dickemann, 1982), is characteristic of the *least* economically successful sectors of the population.   Young males, or males with scant resources, are most predisposed to engage in potentially polygynous but low investment strategies such as rape (Thornhill & Thornhill, 1983; Shields & Shields, 1983; see also Weinrich, 1977) or, as described for 17th-19th century England, sequential common law marriages (Menefee, 1981).   While wealthy men may also pursue polygynous mating strategies, under the legal guise of monogamy (see Essock Vitale, 1984, for U.S. data), the above examples indicate that polygynous marriage or mating is by no means necessarily only the prerogative of the wealthy in all societies.

Secondly, polygyny occurs in societies where, it is maintained, there is no inequality in access to material goods (Chagnon, 1979a).   Among the Yanomamo, for instance, some men have more wives than others on account of their ability to raid other villages for women, and then to protect these wives from competitors.   The size of a man's kinship group, rather than any inequality in access to resources, is the key factor implicated in differences in the competitive abilities of men (Chagnon, 1979b, 1982).   On the basis of these observations, some suggest that Yanomamo polygyny can be characterised as harem defence rather than resource defence polygyny (Flinn & Low, 1986).   This view is difficult to defend in so far as Chagnon's own descriptions illustrate that men do differ according to their ability to utilise resources (Chagnon, 1979a).   In particular, men of large lineages can amass the consumption requirements of polygynous households by means of exploiting the efforts of younger, obligated, dependent kin.   Two interesting points come out of the discussion of polygyny among the Yanomamo.   First, the incidence of polygyny within a particular society may be contingent on differential *utilisation* of resources, rather than differential *access*.   Given the predominance of labour-intensive modes of production in many societies (e.g. Crook & Crook, 1988), this focus on resource utilisation, rather than resource access, is likely to be of general importance in studies of the incidence of polygyny. Secondly, Chagnon suggests that polygyny is only likely to occur in environments where resources are abundant, predictable and require little time and effort to amass; where this is not so, Chagnon suggests "the economic activities of younger kin do not lend themselves readily to the amassing of predictable suprasufficient produce, and the polygynous household is a less viable institution" (Chagnon, 1979a). The occurrence of polygyny in highly labour-intensive societies of Asia (cf. Goody, 1976) suggest this conclusion should be modified.

Thirdly, the type of polygyny described by Dickemann (1979) in highly stratified medieval Chinese, Indian and European elites appears to be contingent on the practice of hypergyny. Parents in middle ranks are eager that their daughters should be married into high ranking families, probably for motives of status, and this gives rise to a system of hypergyny whereby marriageable women, offered with valuable dowries to the sons-in-heir of the elite, accumulate at the top of the social hierarchy. Although the model still requires definition (Kitcher, 1985), Dickemann has suggested an interesting evolutionary route to polygynous marriage. This type of polygyny can still be characterised as resulting from differential resource defence (in so far as social stratification is based on differential resource ownership), but the extent to which the polygyny threshold model is applicable cannot easily be determined with currently available data.

Fourthly, Betzig (1986) offers a further reason why polygyny is found in economically stratified societies. Following Carneiro (1970) and Lee (1979), she argues that the extent to which the members of a society are unable to leave, due to geographic barriers, inhospitable terrain or unfriendly neighbours, will determine the degree of stratification, and therefore of polygyny (cf. Vehrencamp, 1983). Quite simply, if the costs of leaving the group exceed those of staying, a man may be forced into staying in the group, despite the fact that the women of this group are disproportionately monopolised by the wealthy and powerful. Again resource differentials are implicated, but emphasis is placed on the fact that low ranking men have no options but to surrender their women. The relevance of the concept of a polygyny threshold is questionable in situations where wives are accumulated by extortionary means.

Unearthing the ecological bases of human marriage systems has barely begun. Nevertheless, evolutionary biological anthropologists are beginning to examine the general social and economic conditions that determine the nature of

competition between members of one sex for access to members of the other
(see, for example, Crook & Crook, 1988, for polyandry), and as such to provide
some explanation for the diversity of human marriage.

ACKNOWLEDGMENTS

Thanks to Tim Caro for critical reading of an earlier draft; to all
members of the Large Animal Research Group, Department of Zoology,
Cambridge, for their hopsitality and ideas; to the National Geographic Society,
and the Northwestern University Department of Anthropology for funding; and
to the people of Tabarit, Abosi and Kamerumeru for their good-humoured
tolerance of my persistant interest in polygyny.

REFERENCES

Altmann, S.A., Wagner, S.S. & Lenington, S. (1977).   Two models for the
    evolution of polygyny. Behavioral Ecology and Sociobiology **2**, 397-410.

Betzig, L.L. (1986).   Despotism and Differential Reproduction: A Darwinian
    View of History. New York: Aldine.

Blumberg, R.L. & Winch, R.F. (1972).   Societal complexity and familial
    complexity. American Journal of Sociology, **77**, 898-920.

Borgerhoff Mulder, M. (1985).   Polygyny threshold: a Kipsigis case study.
    National Geographic Research Reports, **21**, 33-39.

Borgerhoff Mulder, M. (1987a). On cultural and reproductive success; Kipsigis
    evidence.   American Anthropologist, **89**, 617-634.

Borgerhoff Mulder, M. (1987b).   Resources and reproductive success in women
    with an example from the Kipsigis. J.Zool.Lond., **213**, 489-505.

Borgerhoff Mulder, M. (1988a). Reproductive success in three Kipsigis cohorts.
    In: T. H. Clutton-Brock (ed.),   Reproductive Success: Studies of
    Individual Variation. Cambridge: Cambridge University Press (in press)

Borgerhoff Mulder, M. (1988b).   Kipsigis bridewealth payments.   In:
    L. L. Betzig, M. Borgerhoff Mulder & P. W. Turke (eds.),   Human
    Reproductive Behaviour: A Darwinian Perspective.   Cambridge:
    Cambridge University Press (in press).

Borgerhoff Mulder, M. & Caro, T. M. (1983).   Polygyny: definition and
    application to human data. Animal Behaviour, **31**: 609-610.

Borgerhoff Mulder, M. & Caro, T. M. (1985).   The use of quantitative
    observational techniques in anthropology. Current Anthropology, **26**,
    323-335.

Brown, D. (1988).   Components of lifetime reproductive success.   In: T. H.
    Clutton-Brock (ed.),   Reproductive Success: Studies of Individual
    Variation. Chicago: Chicago University Press (in press).

Bygott, J. D., Betram, B.C.R. & Hanby, J. P. (1979).   Male lions in coalitions
    gain reproductive advantages. Nature, **282**, 839-841.

Caro, T. M. & Collins, D. A.   (1987)   Male cheetah social organization and
    territoriality. Ethology, **74**, 52-64.

Carneiro, R.L. (1970). A theory of the origin of the state. Science, **169**, 733-738.

Central Bureau of Statistics (1980). Kenya Fertility Survey, 1977-78. Nairobi: Central Bureau of Statistics, Ministry of Economic Planning and Development.

Chagnon, N. A. (1979a). Is reproductive success equal in egalitarian societies? In: N. A. Chagnon and W. G. Irons (eds.), Evolutionary Biology and Human Social Behavior: An Anthropological Perspective, pp. 86-132. North Scituate, Mass.: Duxbury Press.

Chagnon, N. A. (1979b). Mate competition, favoring close kin and village fissioning among the Yanomamo Indians. In: N. A. Chagnon and W. G. Irons (eds.), Evolutionary Biology and Human Social Behavior: An Anthropological Perspective, pp. 86-132. North Scituate, Mass.: Duxbury Press.

Chagnon, N.A. (1982). Sociodemographic attributes of nepotism in tribal populations: Man the rule breaker. In: King's College Sociobiology Group (eds.), Current Problems in Sociobiology. Cambridge: Cambridge University Press.

Clignet, R. (1970). Many Wives, Many Powers. Evanston: Northwestern University Press.

Clutton-Brock, T. H., Guinness, F. E. & Albon, S. A. (1982). Red Deer: Behavior and Ecology of Two Sexes. Chicago: Chicago University Press.

Clutton-Brock, T. H. (1983). Selection in relation to sex. In: D. S. Bendall (ed.), Evolution from Molecules to Men. Cambridge: Cambridge University Press.

Clutton-Brock, T. H., Albon, S. A. & Guinness, F. E. (1985). Parental investment and sex differences in juvenile mortality in birds and mammals. Nature, **313**, 131-133.

Clutton-Brock, T. H. (1988). Reproductive success, selection and adaptation. In: T. H. Clutton-Brock (ed.), Reproductive Success: Studies of Individual Variation. Chicago: University of Chicago Press (in press).

Comaroff, J. L. & Comaroff, J. (1980). The management of marriage in a Tswana chiefdom. In: E. J. Krige & J. L. Comaroff (eds.), Essays on African Marriage in Southern African, pp. 24-49. Capetown: Juta.

Crook, J. H. & Crook, S. J. (1988). Tibetan polyandry: problems of adaptation and fitness, In: L. L. Betzig, M. Borgerhoff Mulder & P. W. Turke (eds.), Human Reproductive Behaviour: A Darwinian Perspective. Cambridge: Cambridge University Press (in press).

Daly, M. & Wilson D.(1983). Sex, Evolution and Behavior. Boston: Willard Grant Press.

Dickemann, M. (1979). Female infanticide, reproductive strategies and social stratification: a preliminary model, In: N. A. Chagnon and W. Irons (eds.), Evolutionary Biology and Human Social Behavior: An Anthropological Perspective, pp. 321-367. North Scituate, Mass.: Duxbury Press.

Dickemann, M. (1982). Commentary on Hartung. Current Anthropology, **23**, 1-12.

Downhower, J. F. & Armitage, K. B. (1971). The yellow-bellied marmot and the evolution of polygyny. American Naturalist, **105**, 255-270.

Elliott, P. F. (1975). Longevity and the evolution of polygamy. American Naturalist, **109**: 281-287.

Emlen, S. T. & Oring, L. W. (1977). Ecology, sexual selection and the evolution of mating systems. Science, **197**, 215-223.

Essock-Vitale, S. M. (1984).   The reproductive success of wealthy American, Ethology and Sociobiology, **5**, 45-49.

Evans-Pritchard, E. E. (1940). The Nuer. London: Oxford University Press.

Faux, S. F. & Miller, H. L. Jr. (1984).   Evolutionary speculations on the oligarchic development of Mormon polygyny. Ethology and Sociobiology, **5**, 15-31.

Flinn, M. V. & Low, B. S. (1986).   Resource distribution, social competition and mating patterns in human societies,   In: D. I. Rubenstein & R. W. Wrangham (eds.), Ecological Aspects of Social Evolution, pp. 217-243. Princeton: Princeton University Press.

Fortes, M. (1962).   Introduction in: M. Fortes (ed.),   Marriage in Tribal Societies, pp. 1-13. Cambridge: Cambridge University Press.

Goldschmidt, W. & Kunkel, E. J. (1971).   The structure of the peasant family. American Anthropologist, **73**, 1058-1076.

Goody, J. (1976).   Production and Reproduction: A comparative study of the domestic domain. Cambridge: Cambridge University Press.

Gray, P. J. (1985).   Primate Sociobiology.   New Haven, Connecticut: HRAF Press.

Hamilton, W. D. (1964). The genetical evolution of social behaviour. Journal of Theoretical Biology, **7**, 1-52.

Hartung, J. (1982).   Polygyny and inheritance of wealth. Current Anthropology, **23**, 1-12.

Hartung, J. (1976).   On natural selection and the inheritance of wealth. Current Anthropology, **17**, 607-613.

Hill, J. (1984).   Prestige and reproductive success in man.   Ethology and Sociobiology, **5**, 77-95.

Irons, W. G. (1979a).   Cultural and biological success, In: N. A. Chagnon and W. G. Irons (eds.), Evolutionary Biology and Human Social Behavior: An Anthropological Perspective,   pp. 257-272.   North Scituate, Mass.: Duxbury Press.

Irons, W. (1979b).   Investment and primary social dyads, In: N. A. Chagnon and W. Irons (eds.), Evolutionary Biology and Human Social Behavior: An Anthropological Perspective. North Scituate, Mass.: Duxbury Press.

Irons, E. (1983). Human female reproductive strategies. In: S. J. Wasser (ed.), Social Behavior of Female Vertebrates,   pp. 169-213.   New York: Academic Press.

Kitcher, P. (1985). Vaulting Ambition. Cambridge, Mass.: MIT Press.

Kruijt, J. P. & Vos, G. J. de (1988).   Individual variation in reproductive success in the male black grouse (Tetrao tetrix L.),   In: T. H. Clutton-Brock (ed.), Reproductive Success: Studies of Individual Variation.   Chicago: University of Chicago Press (in press).

Lee, R. B. (1979).   The !Kung San: Men, Women and Work in a Foraging Society. Cambridge: Cambridge University Press.

Le Boeuf, B. J. (1974).   Intra-male competition and reproductive success in elephant seals. American Zoologist, **14**, 163-176.

Le Boeuf, B. J. & Reiter, J. (1988). Lifetime reproductive success in northern elephant seals, In: T. H. Clutton-Brock (ed.), Reproductive Success: Studies of Individual Variation.   Chicago: University of Chicago Press (in press).

Lenington, S. (1980).   Female choice and polygyny in red-winged blackbirds. Animal Behaviour, **28**, 347-361.

Mair, L. (1969). African Marriage and Social Change. London: Cass.

Manners, R. A. (1967). The Kipsigis of Kenya. Culture change in a "model" East African tribe, In: J. Steward (ed.), Contemporary Change in Traditional Societies, vol. 1, pp. 207-359. Urbana: University of Illinois Press.

Menefee, S. P. (1981). Wives for Sale. Oxford: Blackwell.

Orians, G. H. (1969). On the evolution of mating system in birds and mammals. American Naturalist, **103**, 589-603.

Orchardson, I. Q. (1961). The Kipsigis. Nairobi: Kenya Literature Bureau.

Osmand, M. W. (1969). A cross-cultural analysis of family organization. Journal of Marriage and the Family, **31**, 302-310.

Packer, C., Herbst, L., Pusey, A. E., Bygott, J. L., Hanby, J. P., Cairns, S.J. & Borgerhoff Mulder, M. (1988). Reproductive success of lions, In: T. H. Clutton-Brock (ed.), Reproductive Success: Studies of Individual Variation. Chicago: University of Chicago Press (in press).

Peristiany, J. G. (1939).   The Social Institutions of the Kipsigis.   London: Routledge & Kegal Paul.

Peters, E. L. (1980). Aspects of Bedouin bridewealth among camel herders in Cyrenaica. In: J. L. Comaroff, The Meaning of Marriage Payments, pp. 125-159. New York: Academic Press.

Prins, A. H. (1953). East African Age-Class Systems: An Inquiry into the Social Order of the Galla, Kipsigis and Kikuyu. Groningen: The Free Press.

Ralls, K. (1977).   Sexual dimorphism in mammals: Avaian models and unanswered questions. American Naturalist, **111**, 917-938.

Scott, E. C. (1982). Commentary on Hartung. Current Anthropology, **23**, 10.

Shields, W. M. & Shields, L. M. (1983).   Forcible rape: an evolutionary perspective. Ethology and Sociobiology, **4**, 115-136.

Smith, J. E. & Kunz, P. R. (1976). Polygyny and fertility in nineteenth century America. Population Studies, **30**, 465-480.

Spencer, P. (1980). Polygyny as a measure of social differentiation in Africa. In: J. Mitchell (ed.), Numerical Techniques in Social Anthropology, pp. 117-160. Philadelphia: Institute for the Study of Human Issues.

Thornhill, R. & Thornhill, N. (1983). Human rape: an evolutionary analysis. Ethology and Sociobiology, **4**, 137-173.

Thornhill, R. & Alcock, J. (1983). The Evolution of Insect Mating Systems. Cambridge, Mass.: Harvard University Press.

Trivers, R. L. & Willard, D. E. (1973). Natural selection and parental ability to vary sex ratio of offspring. Science, **179**, 190-192.

van den Berghe, P. L. (1979). Human Family Systems: An Evolutionary View. New York: Elsevier.

Vehrencamp, S. L. (1983).   A model for the evolution of despotic versus egalitarian societies. Animal Behaviour, **31**, 667-682.

Verner, J. & Willson, M. F. (1966). The influence of habitats on mating systems of north American passerine birds. Ecology, **47**, 143-147.

Vos, J. G. de (1983). Social behaviour of Black Grouse: An observational and experimental field study. Ardea, **71**, 1-103.

Wade, M. J. & Arnold, S. J. (1980).   The intensity of sexual selection in relation
    to male sexual behaviour, female choice, and sperm precedence.   Animal
    Behaviour, **28**, 446-461.

Weinrich, J. D. (1977).   Human sociobiology: pair-bonding and resource
    predictability (effects of social class and race).   Behavioral Ecology and
    Sociobiology, **2**, 91-118.

Werken, H. van de (1968).   Cheetahs in captivity: preliminary report of
    cheetahs in zoos and in Africa. Zool. Garten, **35**, 156-161.

Wilson, M. (1981).   Xhosa marriage in historical perspective, In: E. J. Krige &
    J. L. Comaroff (eds.), Essays in African Marriage in Southern Africa, pp.
    133-147. Capetown: Juta.

Wittenberger, J. F. (1981).   Male quality and polygyny: the "sexy son"
    hypothesis revisited. American Naturalist, **117**, 329-342.

Yalman, N. (1971).   Under the Bo Tree.   California: University of California.

INDEX